IIW Collection

IIW International Institute of Welding

The IIW Collection of Books is authored by experts from the 59 countries participating in the work of the 23 Technical Working Units of the International Institute of Welding, recognized as the largest worldwide network for welding and allied joining technologies.

The IIW's Mission is to operate as the global body for the science and application of joining technology, providing a forum for networking and knowledge exchange among scientists, researchers and industry.

Published books, Best Practices, Recommendations or Guidelines are the outcome of collaborative work and technical discussions-they are truly validated by the IIW groups of experts in joining, cutting and surface treatment of metallic and non-metallic materials by such processes as welding, brazing, soldering, thermal cutting, thermal spraying, adhesive bonding and microjoining. IIW work also embraces allied fields including quality assurance, non-destructive testing, standardization, inspection, health and safety, education, training, qualification, design and fabrication.

More information about this series at http://www.springer.com/series/13906

Vilia Elena Spiegel-Ciobanu · Luca Costa ·
Wolfgang Zschiesche

Hazardous Substances in Welding and Allied Processes

INTERNATIONAL INSTITUTE OF WELDING
A world of joining experience

Vilia Elena Spiegel-Ciobanu
FB Holz und Metall der DGUV
Berufsgenossenschaft Holz und Metal
Hannover, Niedersachsen, Germany

Luca Costa
Istituto Italiano della Saldatura
Genoa, Italy

Wolfgang Zschiesche
Institut für Prävention und
Arbeitsmedizin der Deutschen
Gesetzlichen Unfallversicherung
Institut der Ruhr-Universität Bochum
Bochum, Nordrhein-Westfalen, Germany

ISSN 2365-435X ISSN 2365-4368 (electronic)
IIW Collection
ISBN 978-3-030-36928-6 ISBN 978-3-030-36926-2 (eBook)
https://doi.org/10.1007/978-3-030-36926-2

This Springer imprint is published by the registered company Springer Nature Switzerland AG
The registered company address is: Gewerbestrasse 11, 6330 Cham, Switzerland

Gas-shielded metal arc welding (MIG/MAG) with extractor, courtesy of Fa. Kemper

Notes

The document is based on the Information of the German Social Accident Insurance (DGUV-Information) 209-017 "Hazardous substances in welding and allied processes," November 2012, Dr.-Ing. Vilia Elena Spiegel-Ciobanu, BGHM, Germany; translation by Karin Hentschel, Annelie Beyer, Berufsgenossenschaft Holz und Metall.

The document has been rearranged by C VIII "Health, Safety and Environment" of the IIW for the scope of international use and reference.

General remark: If not indicated otherwise, tables and figures in this document are taken from this source.

Preface I

Hazardous substances in welding and its allied processes including soldering and brazing consist of particulates and gases. For decades, these are considered as essential items in the identification and the hazard and risk assessment of welding technologies.

Commission VIII "Health, Safety and Environment" of IIW has always kept a close eye on the fume and gas emissions in welding technologies.

In particular, the constituents, the morphology and the structure of welding fume including the question of particular effects of ultrafine particles have been under critical review regarding specific health hazards.

Also, gases that are generated in the course of welding, such as ozone and nitrogen oxides, have been under constant review by Commission VIII.

Recently, new technologies such as laser welding, cutting and additive manufacturing technologies have been added to the scope of "health and safety" as well as organic substances, e.g., emitted by thermal degradation and combustion of coatings such as primers, paints and other protective coatings and also from processing polymers and plastics.

All these aspects have been widely regarded in several IIW Commission VIII documents, the scientific literature, publications of national welding societies and health and safety authorities as well as in national and international standards.

This booklet for the first time compiles the current knowledge on practically all hazardous substances in welding, soldering, brazing and other allied processes that are relevant from health and safety aspects.

It restricts itself to information on process-dependent emission data and general information on health hazards and possible health effects of particles, gases and their possible constituents.

It does not provide information on concentrations of the substances in the work environment and allowable exposure limits as these depend on the individual working situations and on the respective national regulations, both of which are subjects of great variations.

   Taking this into account, this booklet will provide exhaustive and comprehensive information on a wide range of hazardous substances from all major technologies that are relevant in welding and allied processes; the booklet also goes into sufficient detail to provide the reader with information needed for a general hazard analysis.

Bochum, Germany   Wolfgang Zschiesche
                                     On behalf of the authors and Commission VIII

Preface II

Modern development in the welding industry together with the use of new and different materials has created the need for a continuous improvement in the systems and processes needed to protect employees from accidents and workplace illnesses. The intent is for an overall reduction in occupational exposures.

Regulations vary from jurisdiction to jurisdiction depending on focus and so compliance to regulations, while necessary from a legal perspective is not sufficient to ensure a safe working environment. A hazard assessment is necessary to identify the risks, both current and future, to worker safety. A safety program, which provides continual improvement, is necessary to meet the needs of the modern welding workplace. In recent years, changes in health and safety legislation were made, both on a national and on an international level, concerning the reduction of limit values for hazardous substances, e.g., manganese, nitrogen oxides, hexavalent chromium.

The European Chemicals Act has experienced fundamental modifications, which are mainly attributed to two EC regulations:

- The REACH Regulation (edition December 2006) concerning the registration, evaluation, authorization of chemical substances,
- The CLP Regulation (edition January 2009) concerning the classification, labeling and packaging of substances and mixtures.

The Toxic Substances Control Act (TSCA) is a US law, administered by the United States Environmental Protection Agency that regulates the introduction of new or already existing chemicals.

Besides generally applied procedures, further protective measures corresponding to the extent of the hazard are necessary.

This information contained in safety data sheets/product information sheets is of primary importance for hazard assessment.

The following guidelines should be used when determining protective measures:

1. Work should be designed so that hazards to life and health are avoided and the residual risk is kept as low as possible.

2. Hazards should be prevented at source.
3. State of the art, occupational medicine, occupational science and occupational hygiene should be utilized.
4. Engineering, administration, social relations and environmental influences should be managed harmoniously and synergistically.
5. Engineering controls are primary, and personal protection measures are secondary.
6. Employees in need of special protection should be given consideration.
7. Safety and health information should be communicated and managed in a fashion that creates a positive motivational environment for employees.

During welding, cutting and allied processes, gaseous and particulate substances are formed, which, based on composition, concentration and duration of exposure, may present a hazard to the health of the employees (hazardous substances).

To determine the requirements of a successful health and safety program, testing to understand the concentration and intensity of key hazardous substances should be performed.

The purpose of this booklet is:

– to provide information on the generation and the effects of hazardous substances produced during welding and allied processes (thermal cutting, thermal spraying, soldering and brazing etc.),
– to give guidance on the determination of hazardous substances,
– to simplify assessment of the hazard due to hazardous substances.

In addition, the booklet is aimed at suggesting possibilities of how to avoid or reduce the risk to worker's health resulting from exposure to these substances.

Hannover, Germany   Vilia Elena Spiegel-Ciobanu[1]
Genoa, Italy   Luca Costa
Bochum, Germany   Wolfgang Zschiesche

---

[1]affiliated to Berufsgenossenschaft Holz und Metall till End 2017, actually consultant and expert for the German Association for Welding and Allied Processes DVS

Acknowledgements

I express my appreciation and thanks to my co-authors and the other colleagues within Commission VIII:

- Steve Hedrick, especially for the part concerning personal protective equipment (Chap. 8),
- Vincent van der Mee, especially for the contributions on filler wires,
- Nicolas Floros, for the information concerning the welding fume structure,
- John Petkovsek, for providing support and many comments in detail,
- David Jordan, for preparing the part concerning occupational limit values (Sect. 1.5),
- Dave Hisey and Martin Cosgrove, for the contribution to several parts of this booklet.

Many thanks to all the other members of Commission VIII that have helped and provided support along the way to finish this work.

Especially, I want to express my gratitude to the colleagues Annelie Beyer and Karin Hentschel for the English translation of the original German information booklet that was the base of the actual publication.

This booklet is dedicated in memory of Ingo Grothe (once chairperson of Commission VIII), the mentor from the beginning of my occupational way starting at the "Nordwestliche Eisen- und Stahl-Berufsgenossenschaft" in Hannover who introduced me to the international work of Commission VIII, a great family in the field of welding and allied processes.

Vilia Elena Spiegel-Ciobanu

Contents

Chapter 1
General Information on Hazardous Substances in Welding and Allied Processes

1.1 Definition

Hazardous substances in welding and allied processes are the inhalable/respirable substances generated, which are absorbed by the human body (Fig. 1.1). At a certain concentration, these substances may cause a risk to health.

For the scope of this document, welding fume is intended as fume generated by welding and allied processes.

1.2 Classification

Hazardous substances generated during welding, cutting and allied processes can be classified according to their occurrence and effects (Fig. 1.2 and Fig. 1.5).

1.2.1 Occurrence

Hazardous substances are generated by welding, cutting and allied processes in the form of gases and/or particles. Particulate substances are a dispersed distribution of minute solid particles in air. For all particles present in air, the following fractions are distinguished based on particle size (per ISO 7708:1995):

inhalable fraction—fraction of particles which is inhaled through mouth and nose into the body; it comprises particle sizes up to and exceeding 100 μm. In the past, this fraction was called "total dust".

respirable fraction—fraction of particles capable of penetrating into the alveoli (air sacs); it comprises particle sizes up to 10 μm. In the past, this fraction was called "fine dust".

© International Institute of Welding 2020
V. E. Spiegel-Ciobanu et al., *Hazardous Substances in Welding and Allied Processes*, IIW Collection, https://doi.org/10.1007/978-3-030-36926-2_1

Fig. 1.1 Absorption of hazardous substances by the human body by inhalation. *Source* [1]. After Ruwac-Industries-Exhauster-GmbH, 1993

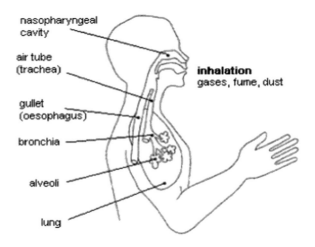

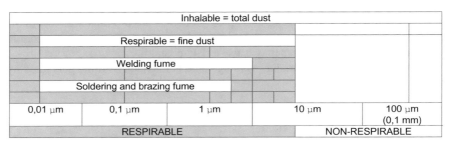

Fig. 1.2 Classification of particulate hazardous substances/welding fume in welding and allied processes according to their particle size (occurrence). *Source* [1]

Particulate substances generated during welding are very fine. In general, they have a diameter of less than 1 μm (in most cases less than 0.1 μm), therefore they are respirable and called "welding fume". Particles in the size range of <0.1 μm are called "ultrafine particles".

During thermal cutting and some allied processes, the particulate substances generated are only partially respirable.

Particle size, structure and morphology (shape)

The quantity of particles depends on the combination of the processes and materials used.

Depending on the process group, different particle sizes with different particle morphology result (Fig. 1.3).

Morphological studies suggest that the individual welding fume particles do not have a homogeneous composition.

Process	Material	Shape of individual particles	Particle		
			Size of		
			Individual particles (diameter)	Chains (length)	Agglomerates (diameter)
Manual metal arc welding with covered electrodes (MMA)	Cr-Ni steel	ball shaped	up to 50 nm	several μm	up to 500 nm
			up to 400 nm	several μm	
Gas shielded arc welding (MAG/MIG)	Cr-Ni steel	ball shaped	up to 10 nm	up to 100 nm	up to 100 nm
Gas shielded arc welding (MIG)	Aluminium alloys	ball shaped	10 to 50 nm	n.d.	
			up to 400 nm	n.d.	n.d.

n.d. = no data
μm = micro metres (1 μm = 10^{-3} mm = 10^{-6} m)
nm = nano metres (1 nm = 10^{-6} mm = 10^{-9} m)

Fig. 1.3 Particle size, shape and morphology of welding fume (examples). *Source* Spiegel-Ciobanu

Structure

The morphology of the particles may play an important part in determining the absorbability and thus toxicological potential in welding fumes. Spherical particles in welding fume, especially the larger ones, may have a core and shell [1–4]. Cores generally consist of a complex of iron-rich oxides containing iron, manganese, potassium, chromium, nickel and other transition elements and oxygen in different ratios depending on the type of consumable. The metal oxide core of these particles is surrounded by a relatively impervious shell rich in amorphous silicon oxide or silicates containing other compounds such as iron and manganese fluoride, (silicon oxides with other elements like natrium, kalium, manganese) depending on the composition of the flux of the wire.

Besides primary particles (individual particles) chains and agglomerates are also formed by coagulation (Figs. 1.4a, b).

(a)

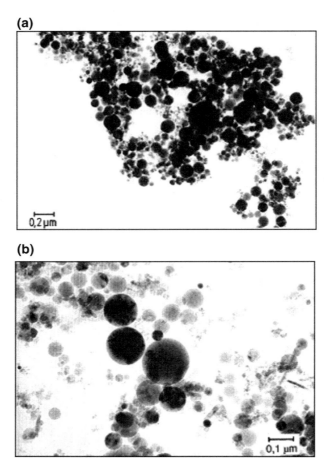

(b)

Fig. 1.4 **a** Particles of the fume generated during metal inert gas welding under carbon dioxide. *Source* Courtesy Welding and Joining Institute, Technical University Aachen. **b** Electron microscope photos of welding fume—Particles of the fume generated during metal inert gas welding of aluminium alloys. *Source* Courtesy Welding and Joining Institute, Technical University Aachen

1.2.2   Health Effects

The gaseous and particulate substances generated during welding, cutting and allied processes can be classified according to their effects on the different sites and organs of the human body as in Fig. 1.5.

Substances stressing the respiratory tract and the lung—long-term intake of high concentrations may lead to stress of the respiratory tract and the lung. A long-term intake of high concentrations may e.g. lead to diseases of the respiratory tract (in the form of chronic bronchitis, obstructive airways' disease and chronic obstructive lung disease (COPD)).

Occurrence			Health Effects		
gaseous	Particulate[1]		stressing to airways and/or lungs	Irritant/toxic/ fibrogenic	carcino- genic
	inhalable	respirable			
nitrogen monoxide				X	
nitrogen dioxide				X	X[2]
ozone				X	X[2]
carbon monoxide				X	
phosgene				X	
hydrogen cyanide				X	
formalde- hyde				X	X[2]
	aluminium oxide		X	X[3]	
	iron oxide		X	X[3]	
	magnesium oxide		X		
	calcium oxide (from basic coated or cored consumables)		X	X	
	barium compounds			X	
	lead oxide			X	
	fluorides			X	
	copper oxide			X	
	zinc oxide			X	
	manganese oxide			X	
	molybdenum oxide			X	
	chromium VI compounds			X	X
	nickel oxide			X	X
	cobalt oxide			X	X[2]
	cadmium oxide			X	X
	beryllium oxide			X	X

1) Notice that some compounds, in particular metal oxides, are not necessarily formed as single oxides, the effects of which are described here. They may be part of a complex welding fume matrix, such as spinells, which may behave differently in the the organism.
2) Suspected to have a carcinogenic effect
3) Aluminosis and siderosis/siderofibrosis resp. in rare cases

Fig. 1.5 Classification of hazardous substances according to their occurrence and health effects. *Source* [1]

In addition, dust deposits in the lung may occur as siderosis (for iron oxides). Furthermore, at high concentrations, fibrogenous reactions (reproduction of the connective tissue) of the lung may occur (e.g. for aluminium oxide).

Toxic (poisonous) substances—have a toxic effect on the human body, if a certain concentration or a certain dose (concentration x time) is exceeded. There is a concentration/dose-effect-relationship. This includes also irritant effects to the airways and the lungs. Some pollutants may also lead to metal fume fever, such as zinc oxide and copper oxide.

Carcinogenic (cancer causing) substances—are substances which are known to cause malignant tumors or other malignancies. The latency period (interval between first or last exposure and the outbreak of the disease) may last for years or decades.

For these substances, there is mostly no known threshold value below which the health hazard does not exist. In many cases these substances have an additional toxic effect.

Carcinogenic substances are listed e.g. in the "Amendment of the Regulation EC No. 1272/2008 of the European Parliament and of the Council on classification, labelling and packaging of substances and mixtures", by the US National Toxicology Program 14th Report on Carcinogens and by the International Agency for Research on Cancer (IARC) of the WHO (for welding fumes: see [5]). A more differentiated scheme on the carcinogenicity of substances has been elaborated by the German Research Foundation (DFG) [6].

1.3   Generation

Hazardous substances generated during welding and allied processes arise from:

- filler materials
- parent materials
- shielding gases
- coatings
- contamination and
- ambient air/neighbourhood

at high temperature (of the arc or flame) by physical and/or chemical processes (Fig. 1.6) such as

- evaporation
- condensation
- oxidation
- decomposition
- pyrolysis and
- combustion.

The type and amount of the hazardous substances generated depend on the material and the process. The chemical composition of the materials used has a direct influence on the chemical composition of the particulate hazardous substances. The processes used affect the generation of gaseous hazardous substances.

Evaporation	metals	
	$\left.\begin{array}{l}\\ \\ \end{array}\right\}$	Fe, Cu, Mn, Ni, ..
Condensation	metals	
Oxidation	metals + O_2 = oxides	FeO, Fe_2O_3, CuO, ...
	$N_2 + O_2 \rightarrow 2NO$	
	$NO + 1/2O_2 \rightarrow NO_2$	
Decomposition	$CO_2 \rightarrow CO + 1/2\ O_2$	
Pyrolysis	organic components	
	$C_xH_y \rightarrow C_{x1}H_{y1}$	
	CO	
	CH_2O	
Combustion	organic components + O_2	
	O_2	
	$C_xH_y \rightarrow CO + H_2O$	
	$CO_2 + H_2O$	

Fig. 1.6 Generation of hazardous substances (examples). *Source* Spiegel-Ciobanu

1.3.1   Gaseous Hazardous Substances

Carbon monoxide (CO) is generated in critical concentrations during metal active gas welding with carbon dioxide (MAGC) or during metal active gas welding with mixed gases (with a high concentration of carbon dioxide) by thermal decomposition of carbon dioxide (CO_2).

Furthermore, carbon monoxide is generated during any form of combustion with an inadequate oxygen supply.

Nitrogen oxides (NO_x = NO, NO_2) are generated by oxidation of the atmospheric nitrogen (from the oxygen [O_2] and nitrogen [N_2] in the air) at the edge of the flame or the arc. Nitrogen monoxide is generated at temperatures exceeding 1000 °C. Nitrogen monoxide oxidises to nitrogen dioxide in the air at room temperature.

$$N_2 + O_2 \xrightarrow{T > 1000°C} 2\,NO$$

$$2\,NO + O_2 \xrightarrow{T_{Room}} 2\,NO_2$$

In oxy-fuel processes (gas welding, flame heating, flame straightening, flame cutting, flame spraying), in plasma cutting with compressed air or nitrogen, and in laser beam cutting with compressed air or nitrogen, the predominant hazardous substances (key components) are nitrogen oxides (primarily nitrogen dioxide).

Ozone (O_3) is generated by ultraviolet radiation from the oxygen in the air, especially during inert gas shielded welding of materials reflecting radiation strongly, such as aluminium and aluminium silicon alloys. The presence of other gases, fume or dust in the air accelerates the decomposition of ozone into oxygen.

$$O_2 \xrightarrow{UV\ radiation} 2O$$

$$O + O_2 \rightarrow O_3$$

$$O_3 \xrightarrow{gas\ and\ particles} O_2 + O$$

This explains why the ozone concentration is particularly high in processes with low fume generation.

Phosgene ($COCl_2$) is generated during heating or by UV-radiation of degreasing agents containing chlorinated hydrocarbons.

Gases from coating materials are generated during welding of workpieces with shop primers (surface coatings preventing corrosion) or with other coatings (paints, lacquers). Depending on the chemical composition of the coatings, not only metal oxides are generated, which are particulate, but also gases, e.g. carbon monoxide (CO), formaldehyde (HCHO), isocyanates, hydrogen cyanide (HCN), hydrogen chloride (HCl).

1.3.2 Particulate Hazardous Substances

Due to the complex thermodynamic conditions during formation of the welding fume, a chemical element may form different compounds (oxides, mixed oxides/complex oxides/spinel) in the welding fume, as a function of the proportion of this element and the presence of other chemical elements in the welding fume [7–10].

Such examples are: Fe_3O_4, $MnFe_2O_4$, K_3FeF_6, CaF_2 etc.

Because of the complex chemical composition and complex structure of the particulate hazardous substances in welding fume and the proportion of each component and chemical compound in welding fume that depends strongly on a lot of different parameters, it's appropriate for the practice (measurements at the workplace, risk assessment) and at the same time necessary to simplify and to use simple oxide instead of spinel. Also, the existing limit values generally don't refer to spinel or complex elements oxides.

The further information gives an overview about the different possible oxides and other chemical compounds that are generated during welding and allied processes.

Iron oxides (FeO, Fe$_2$O$_3$, Fe$_3$O$_4$) (Spinel-type compounds: MnFe$_2$O$_4$, CaFeSi$_2$O$_6$) are generated from filler and parent material during welding and cutting of steels.

Aluminium oxide (AI$_2$O$_3$) is generated from filler and parent material during welding and cutting of aluminium base materials.

Manganese oxides (MnO$_2$, Mn$_2$O$_3$, Mn$_3$O$_4$, MnO) (Spinel-type compounds: MnFe$_2$O$_4$, MnSiO$_3$, MnTiO$_3$) are generated by any arc process using manganese containing fillers. The concentration of manganese in the welding filler has a direct influence on the concentration of manganese oxide in the welding fume and always leads to enrichment in the welding fume. Analyses during hard facing with core wires having high manganese content revealed emission levels of up to 40% manganese oxides in the welding fume [8, 9].

Fluorides (CaF$_2$, KF, NaF, KCaF$_3$, K$_3$FeF$_6$, K$_2$SiF$_6$, other) are generated from the covering of stick electrodes or from the filling of flux-cored wires when using lime-type coatings or fluxes containing fluorides.

In manual metal arc welding with basic covered unalloyed and low-alloy electrodes, for example, the concentration of fluorides in the welding fume reaches values between 10 and 20%.

Barium compounds (BaCO$_3$, BaF$_2$) are generated during welding with filler materials containing barium from the coating of covered electrodes or from the filling of the flux-cored wires, e.g.:

- electrodes for welding of cast iron and copper alloys,
- high and medium-alloy flux-cored wire electrodes or covered electrodes.

Thus, for example, during welding of cast iron and copper alloys with covered electrodes, the barium content in welding fume reached 40%.

Potassium oxide, sodium oxide, titanium dioxide (K$_2$O, Na$_2$O, TiO$_2$) (Complex compounds: KCaF$_3$, K$_3$FeF$_6$, K$_2$SiF$_6$, NaFeO$_2$) are generated from the coating when covered electrodes are used. Titanium dioxide may also occur in the fume of rutile-acid electrodes.

Chromium (III) compounds (Cr$_2$O$_3$, FeCr$_2$O$_4$, KCrF$_4$) Trivalent chromium compounds are generated in small concentrations in metal arc welding using high alloy covered electrodes. Spinel-type compounds (mixed oxides/complex oxides) as (FeMnNi) (FeMnCr)$_2$O$_4$ and Cr$_2$O$_3$ are in high concentrations generated up to 90% from total chromium in welding fume from MAG-Welding with high alloy wires. They were found also in the welding fume from high-alloyed flux-cored wires (up to 40% from total chromium).

Chromium (VI) compounds (chromates = Na$_2$CrO$_4$, K$_2$CrO$_4$, ZnCrO$_4$, etc.) (chromium trioxide = CrO$_3$) Hexavalent chromium compounds are generated in critical concentrations when using high-alloy covered electrodes for manual metal arc welding and when welding with high-alloy flux-cored wires containing chromium.

Chromium(VI) compounds may also occur in repair welding of materials coated with shop primers containing zinc chromates, which were common in the past.

Nickel oxides (NiO, NiO_2, Ni_2O_3) are mainly generated by:

- welding with pure nickel and nickel-base alloys (from the filler material)
- plasma cutting of high-alloy steel containing nickel (from the parent material)
- thermal spraying with nickel-base spraying materials (from the spraying material).

Cadmium oxide (CdO) is generated:

- from the filler material when brazing with brazing alloys containing cadmium,
- during welding and cutting of cadmium coated material.

Beryllium oxide (BeO) is generated from the parent material used in cutting of material containing beryllium.

Cobalt oxide (CoO) is generated from the:

- filler material used in weld surfacing with alloys containing cobalt,
- spraying material in thermal spraying with cobalt-containing alloys,
- parent material in cutting of steel containing cobalt as an alloying element.

Thorium dioxide (ThO_2) is generated from the thoriated tungsten electrode mainly during TIG welding of aluminium material.

Other metals in the oxide form

Lead oxide (PbO), copper oxide (CuO), zinc oxide (ZnO), tin oxide (SnO) are generated during processing and manufacturing welding operations (e.g. from metallic coatings during repair welding, from the spraying material during thermal spraying, from the flux/filler material during soldering and brazing) using materials containing the above metals.

Fumes from coating materials

A great number of hazardous substances consisting of organic components are generated by welding and cutting processes from metallic materials having organic base coatings (e.g. paints, lacquers, primers).

Also during welding on organic layers as mineral oil from sheet production/milling, plastic coatings, oil and grease (e.g. in repair welding), organic components are generated.

1.3.3 Hazardous Substances from Organic Based Coating Materials

Studies using pyrolysis of organic coatings used in shipbuilding, which are partly still applied today, revealed the decomposition products as shown in Figs. 1.7 and 1.8.

Decomposition products (Hazardous substances)	Key and principal components[1] for different coating materials							
	Intermediate coat (binder base)						Finishing coat[2] (binder base)	
	Shop primer[3]			Primer[4]				
	epoxy resin	ethyl silicate	PVB	epoxy resin	chlorinated rubber	alkyd resin	chlorinated rubber	alkyd resin
aliphatic aldehydes[5]			L_2		L_3	L_4	L_3	L_4
Aliphatic alcohols $(C_2\text{-}C_4)$[6]	L_4	L_2	L_3	L_4				
Aliphatic carboxylic acids						L_2		L_2
Alkyle benzenes $(C_7\text{-}C_8)$[7]	L_3				L_3	L_4	L_4	
Hydrogen chloride (HCl)					L_1		L_1	
Carbon monoxide (CO)	L_2	L_1	L_1	L_2	L_2	L_3	L_2	L_3
Phenols (incl. bisphenol A)	L_1			L_1				
Phthalic Anhydride						L_1		L_1
fine dust (respir. fraction)	L	L	L	L	L	L	L	L

[1] L: general key component. L_1, L_2, L_3, L_4: key component and principal component.
[2] Top coat is also called "finishing paint".
[3] The intermediate coat which the manufacturer often applies on semi-finished products (tools, profiles) is called shop primer.
[4] The intermediate coat which is applied on finished products by the process operator is called primer.
[5] e.g. butyric aldehyde.
[6] e.g. butanol
[7] e.g. toluene, xylene

Note! From experience, it is known that with increasing temperature, the spectrum of decomposition products moves towards low-molecular materials, e.g.:

aliphatic aldehydes	$\Rightarrow$	acrolein, formaldehyde
aliphatic alcohols	$\Rightarrow$	ethanol, methanol
aliphatic carboxylic acids	$\Rightarrow$	acetic acid, formic acid.

Fig. 1.7 Key components related to decomposition products of organic based coating materials during pyrolysis (t = 350 °C). *Source* Spiegel-Ciobanu based on [11]

Coating (binder base)	Epoxy tar amide addition compound hardened	Poly-urethane Tar	Epoxy tar amine hardened	Epoxy resin	Urethane alkyd resin	Epoxy tar	Alkyd resin	Vinyl/epoxy resin (tar containing)
Decomposition products (Hazardous substances)	Detected substances			Key components				
Acenaphthene	X							
Acetaldehydel								X
Benzaldehyde			X					
Benzene	X	X	X	X	X	X	X	X
Biphenyl	X							
Bisphenol-A	X		X	X		X		
Butene								X
4-tert, Butylphenol			X					
Dibenzofuran	X							
Dihydrobenzopyrane or isomers	X							
Diisocyanatetoluene		X						
Fluorene	X							
Cresols			X					
Methyl-methacrylate								X
α-Methylstyrene (Isopropenylbenzene)			X	X		X		
Dimers of α-Methylstyrene								
Methylnaphthaline	X							
Naphthaline	X							
4-core poly aromatic hydrocarbons (PAH)	X	X						
5-core polyaromatic hydrocarbons	X	X						
Phenantrene /Anthracene	X	X						
Phenol	X		X	X				X
Pyrene	X	X						
Styrene	X		X	X	X	X	X	X
Toluene	X	X	X	X	X	X	X	X
Xylene	X		X					

Fig. 1.8 Decomposition products from organic based coating materials during pyrolysis (t = 800 °C). *Source* Spiegel-Ciobanu based on studies of the German Liability Accident Insurance Institution for the Wood and Metal Sectors (BGHM)

1.4 Influencing Factors

Apart from the processes and materials used, the amount and kind of hazardous substances are also influenced by surface coatings and contaminations as well as by the following factors (Fig. 1.9):

- **Current, voltage**: For identical processes and materials, higher welding currents and welding voltages generally lead to higher emission rates.
- **Type of current**: Higher emission rates are generally observed with a.c. than with d.c.

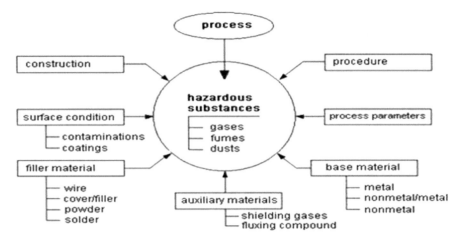

Fig. 1.9 Influencing factors. *Source* Spiegel-Ciobanu

– **Diameter of the electrode**: Emission of welding fume increases with the electrode diameter.
– **Type of coating**: Rutile coated electrodes have the lowest emission rates of welding fume while cellulose covered electrodes have the highest.
– **Inclination angle of the electrode**: At flat angles of inclination of the electrode, emission rates are lower than at steep inclination angles.

In general, it should be considered that parameters should be properly optimized and specified to ensure proper repeatability of conditions.

1.5 Occupational Exposure Limit Values

Occupational exposure limit values (OELs) can be defined as 'the limit of the time-weighted average of the concentration of a chemical agent in the air within the breathing zone of a worker in relation to a specified reference period' [1]. The OEL represents the maximum concentration of a substance that is not expected to induce acute or chronic detrimental health effects. Such limits are defined by national regulatory authorities and generally have statutory force. Some countries have adopted the Threshold Limit Values (TLV) recommended as guidelines (but not as legal standards) by the non-governmental body the American Conference of Governmental Industrial Hygienists. In the European Union, indicative occupational exposure limit values (IOELV) have been defined for certain chemical agents and shall be taken into consideration by member states when they establish national occupational exposure limits.

While occupational exposure limits are sometimes specified for total welding fume, limits for individual components of the fume (e.g. trivalent or hexavalent chromium) shall be considered and are frequently the controlling factor in decid-ing the permissible conditions in the workplace. It should be noted that operations allied to welding, such as grinding of workpieces or dressing of electrodes used for TIG (GTAW) welding, may also generate particles of substances subject to control and must be incorporated in the assessment of working conditions. Occupational exposure limits are systematically reviewed by the regulatory authorities and may be amended to reflect the findings of current medical research. Although limits for specific substances are generally applicable and not restricted to their occurrence in welding operations, the feasibility of applying such limits in practice is some-times considered, when they are defined. In some jurisdictions, the physical form of substances is also considered, general exposure limits differing for inhalable and respirable fractions of particulates.

1.6 Biological Limit Values

Biological limit can be defined as a limit value for the concentration of a substance, its metabolite (transformation product within the body) or a stress indicator in the corresponding biological material on a toxicological and occupational medical basis at which the health of an employee is not impaired. It is a guideline for control of risks since, due to biological variability, an individual's measurement may exceed the limit without incurring an increased health risk. It is nevertheless useful for guidance when monitoring the health of workers. Such limits are specified in several countries. In this context, ACGIH in the USA defines biological exposure indices (BEI) as indicators. While in the European Community, binding biological limit values may be drawn up and, taking into account feasibility factors, limit values established by member states must not exceed them. Where it is not possible to define a limit value, a health-based guidance level may be substituted.

1.7 Test Methods

Reliable evidence of the risk which the welder encounters at the workplace from exposure to hazardous substances can be obtained from application of different test methods, primarily:

– laboratory emission measurements
– workplace exposure measurements—measurements of concentrations
– analyses of biological material
– epidemiological studies.

Laboratory emission measurements determine the amount per unit time (mg/s) and the chemical composition of emitted hazardous substances for several processes and materials using the fume box method; this should be done using appropriate methods as per reference standards, such as ISO 15011-1 or AWS F1.2.

Thus, basic data are obtained for comparison of different processes and materials and for the evaluation of the hazard to the welder (see clause 5). Emission measurements also provide a basis for calculating ventilation systems and for other necessary protective measures.

Workplace exposure measurements are intended to show the external exposure of the welder. Sampling takes place according to ISO 10882-1 in the welder's breathing zone (personal air sampling). Details on the quantitative and qualitative evaluation of the sample are described in clause 6.1 (Measuring methods for gaseous substances) and in clause 6.2 (Measuring methods for particulate substances). **The measured concentrations** (mg/m^3) are compared with the specific limit values and determine the protective measures to be taken. The accuracy of the measurement primarily depends on whether the sampling is actually taken within the breathing zone. There are several approaches in this field at present, with corresponding specifications for measuring techniques (sampling).

Analyses of biological material, i.e. body fluids (urine, blood) taken from the welder, show the concentrations of critical substances they contain. These values give information on the level of the welder's internal exposure caused by exposure at the workplace and are compared with reference values or with Biological Exposure Indices (BEI). The time of taking the specimens (before or after shift, uninterrupted period of work days) must be considered. Any contamination (e.g. from work clothing) must be avoided [6, 12].

Epidemiological studies are carried out to clarify the frequency of diseases and mortality in different groups of persons, e.g. to clarify the welder's pulmonary cancer risk. Epidemiological studies are based on comparisons between "test persons (probands)" (e.g. welders) and a control group (employees not involved in welding activities and hence considered to be unexposed).
Many epidemiological studies have been carried out with welders.
In general, there is an excess rate in arc welders for chronic bronchitis, irrespective of the smoking status and the fume compounds. Quite a few studies also show an increase of obstructive airways' disease and chronic obstructive lung disease (COPD), despite the effects are often small [13].
Many studies show a slight excess risk of lung cancer in arc welders, irrespective of the fume compounds, which can probably not be explained by smoking habits or confounding exposures like asbestos alone [5, 14].

References

1. Spiegel-Ciobanu, V. E. *Hazardous substances in welding and allied processes*. German Social Accident Insurance (DGUV), Information 209-017.
2. Eichhorn, F., Trösken, F., & Oldenburg, Th. (1981). *Untersuchung der Entstehung gesundheitsgefährdender Schweißrauche beim Lichtbogenhandschweißen und Schutzgasschweißen*. Düsseldorf: DVS-Verlag.

3. Eichhorn, F., & Oldenburg, Th. *Untersuchung der Schweißrauchentstehung beim Lichtbogen-handschweißen und beim Schutzgasschweißen mit mittel- und hochle-gierten Zusatzwerkstoffen*. Düsseldorf: DVS-Verlag.
4. Eichhorn, F., & Oldenburg, Th. (1983). *Vergleichende Untersuchung neuerer Schweißverfahren für das Verbindungsschweißen vo Aluminium und seinen Legierungen zum Zwecke der Reduzierung der Schadstoffbelastung des Aluminiumschweißers*. Düsseldorf: DVS-Verlag.
5. International Agency for Research on Cancer (IARC). (2018). *IARC monographs on the evaluation of carcinogenic risks to humans, Volume 118: Welding, molybdenum trioxide, and indium tin oxide*. Lyon: IARC. https://monographs.iarc.fr/.
6. German Research Foundation (DFG). *List of MAK and BAT values 2018*. Wiley Online Library: https://onlinelibrary.wiley.com/doi/book/10.1002/9783527818402 and http://www.dfg.de/dfg_profil/gremien/senat/arbeitsstoffe/.
7. Voitkevich, V. (1995). *Welding fumes: formation, properties and biological effects*. Cambridge, England: Abington.
8. Spiegel-Ciobanu, V. E., Brand, P., Lenz, K., et al. (2014). Characterisation of the biological effect of ultrafine particles in welding fumes after controlled exposure—Effect of the MIG welding of aluminium and the MIG brazing of zinc-coated materials. *Welding and Cutting, 13*(3), 171–176.
9. Spiegel-Ciobanu, V. E. (2006). Parkinson's disease and exposure to manganese during welding. *Welding and Cutting, 5*(2), 106–111.
10. Floros, N. (2018). Welding fumes main compounds and structure. *Welding in the World, 62*, 311–316.
11. Engström, B., Henricks-Eckerman, M.-L., & Ånäs, E. (1990). Exposure to paint degradation products when welding, flame cutting, or straightening painted steel. *American Industrial Hygiene Association Journal, 51*(10), 561–565.
12. American Conference of Governmental Industrial Hygienists (ACGIH): Biological Exposure Indices (BEI®). http://www.acgih.org/tlv-bei-guidelines/biological-exposure-indices-introduction.
13. Cosgrove, M. (2015). Arc welding and airway disease. *Welding in the World, 59*, 1–7.
14. Commission VIII. (2011). "Health, Safety and Environment" of IIW: Lung cancer and arc welding of steels. *Welding in the World, 55*, 12–20.

Chapter 2
Health Effects of Specific Hazardous Substances

The effects can be considered on the basis of the structure of welding fumes (see also Sect. 2.2), as they are gaseous or in the form of particulate matters.

2.1 Toxic Gaseous Hazardous Substances

Figure 2.1 reports possible effects of specific gaseous hazardous substances in welding and allied processes. Some of these substances may be found in very specific welding operations only.

2.1.1 Carbon Monoxide (CO)

Very poisonous, odourless gas. In higher concentrations, the oxygen transport in the blood is impeded by the great affinity of carbon monoxide to haemoglobin (haemoglobin is necessary for transporting oxygen in the body). The result is a lack of oxygen in the tissues.

Dizziness, lassitude, nausea and head ache may occur in early stages of a CO intoxication, in severe cases unconsciousness, coma and death may result. Concerning pregnancy, CO poses a risk to the fetus even within the occupational exposure limits for workers.

2.1.2 Nitrogen Oxides ($NO_x = NO, NO_2$)

Also called nitric oxides or nitrous gases. Nitrogen monoxide (NO) is a colourless, poisonous gas. Nitrogen dioxide (NO_2) is a brown-red, poisonous gas causing oxidation. Nitrogen dioxide is more toxic than nitrogen monoxide and acts as an

© International Institute of Welding 2020
V. E. Spiegel-Ciobanu et al., *Hazardous Substances in Welding and Allied Processes*, IIW Collection, https://doi.org/10.1007/978-3-030-36926-2_2

Hazardous substance		Effect
Toxic		
Nitrogen monoxide (NO)		irritation of the mucous membrane irritant gas intoxication delayed pulmonary edema (life threatening)
Hydrogen cyanide	(HCN)	toxic - impedes oxygen metabolism of the cells poisoning life threatening
Carbon monoxide	(CO)	toxic - impedes oxygen transport in the blood headache poisoning possibly unconsciousness respiratory paralysis
Carcinogenic		
Aldehydes, e. g. Formaldehyde (CH$_2$O)		Suspicion of carcinogenic effect - strong irritation of the mucous membrane
Ozone	(O$_3$)	Suspicion of carcinogenic effect - toxic - irritation of the mucous membrane acute irritant gas intoxication pulmonary edema
Nitrogen dioxide (NO$_2$)		Suspicion of carcinogenic effect - toxic - irritation of the mucous membrane irritant gas intoxication
Phosgene (Carbonylchloride)	(COCl$_2$)	delayed pulmonary edema (life threatening)
Isocyanates (e. g. TDI)		irritation of respiratory tract or immunological effect (bronchial asthma, pneumonitis)

Fig. 2.1 Possible effects of specific gaseous hazardous substances in welding and allied processes. Amended from [1]

insidious irritant gas even in relatively low concentrations. The first stage of intoxication comprises irritation of the air passages and dyspnoea and can be followed by an asymptomatic state lasting from some hours to several days. The second stage leads, in severe cases, to fatal lung edema (accumulation of fluid in the lungs) [2, 3].

2.1.3   Ozone (O₃)

In high concentrations, this is a deep blue gas having a penetrating smell and being highly toxic. It acts as an irritant gas on respiratory organs and eyes. It causes an irritation of the throat, dyspnoea and possibly lung edema.

More recent studies do not exclude the possibility that ozone has a carcinogenic potential.

2.1.4 Phosgene (COCl₂)

(Carbonyl chloride or carbon dichloride oxide)—is an colourless, extremely poisonous gas with a musty smell. Initially (3–8 h) there are only slight symptoms which may be followed by heavy irritations of the respiratory tract ending in lung edema (accumulation of fluid in the lungs).

2.1.5 Gases from Coating Materials and Fluxes

Hydrogen cyanide (HCN) (hydrocyanic acid) and cyanides (the salts of hydrocyanic acid)—have a smell of bitter almonds and are among the strongest and quickest-acting poisons. They impede the oxidative processes within the cells, so that the cells cannot make use from the oxygen Poisoning may be lethal within short time, so that antidotes must be at hand in case there is a chance of cyanide exposure in some conditions of welding, soldering and brazing.

Aldehydes, i.e. Formaldehyde (CH₂O)—are pungently smelling colourless gases having a strongly irritant effect on the mucous membranes. They cause inflammation of the respiratory tract and some of them are suspected to be mutagenic and carcinogenic.

Isocyanates, e.g. Toluylene diisocyanate (TDI)—have a strong irritant effect on the respiratory tract and may be allergenic; they may cause asthma-like attacks and may lead to "bronchial asthma" by sensitisation, in rare cases also to hypersensitivity pneumonitis.

2.2 Particulate Hazardous Substances

Figure 2.2 reports possible effects of specific particulate hazardous substances in welding and allied processes. Some of these substances may be found in very specific welding operations only.

2.2.1 Lung-Stressing Substances

Iron oxides (FeO, Fe₂O₃, Fe₃O₄)—are considered to be substances without any toxic or carcinogenic effects. Long-term intake of high concentrations may result in a dust deposit in the lungs. This deposit is also known as siderotic pneumoconiosis or siderosis. It is also called "tattooing of the lungs". After long-term exposure to high

Hazardous substance	Effect
Lung stressing	
Welding fume, in general	dust deposits in the lung
Aluminium oxide	dust deposits in the lung, aluminosis, lung fibrosis
Iron oxide	dust deposits in the lung, siderosis, siderofibrosis
Toxic	

Barium compounds, soluble	toxic - nausea
	          possible potassium deficiency
	          neuro- and musculotoxicity

Fluorides	toxic - irritation of the mucous membrane bone damage (fluorosis)
Copper oxide	toxic - suspect of metal fume fever (copper fume fever)
Zinc oxide	toxic - metal fume fever (zinc fume fever)
Manganese oxide	toxic - irritation of the mucous membrane damages of the central nervous system
Carcinogenic	
Beryllium oxide	carcinogenic - Acute and chronic beryllium disease (berylliosis)

Lead oxide	suspicion of carcinogenic effect
	toxic - nausea
	      - anemia
	          gastrointestinal disorders
	          nerve and kidney damage

Cadmium oxide	carcinogenic
	      - irritation of the mucous membrane
	- toxic lung edema (fluid in the lungs)
	          pulmonary emphysema
	- toxic kidney damage

Chromium(VI) compounds	carcinogenic (respiratory tract) - irritation of the mucous membrane
Cobalt oxide	suspected carcinogenic - damage to breathing organs
Nickel oxides	carcinogenic (respiratory organs) - irritant to lung and airways
Radioactive	

Thorium dioxide	radioactive
	      - radiation of the bronchi and the lungs can have
	          carcinogenic effects

Fig. 2.2 Possible effects of specific particulate hazardous substances in welding and allied processes. Amended from [1]

concentrations, this particle deposition may result in a lung fibrosis (siderofibrosis) in rare cases [4].

Aluminium oxide (Al_2O_3)—may lead to a dust deposit in the lungs. Under certain circumstances an aluminosis (pneumoconiosis, lung fibrosis) may occur Irritation of the respiratory tract may also occur [5].

Potassium oxide, sodium oxide, titanium dioxide (K_2O, Na_2O, TiO_2)—should be classified as lung-stressing, because they may result in dust deposits in the lungs.

2.2.2 Toxic Substances

Manganese oxides (MnO_2, Mn_2O_3, Mn_3O_4, MnO) in high concentrations may have irritant effects on the respiratory tract and may result in pneumonitis Long-term exposures can impair the central nervous system [6–9].

Fluorides (CaF_2, KF, NaF and others)—high concentrations lead to irritation of the respiratory tract. After long-term exposure to high concentrations bone damages (fluorosis) are observed (unknown in welders up to now).

Barium compounds ($BaCO_3$, BaF_2)—in the welding fume are mainly present in water-soluble form and have a toxic effect when taken in by the human body. In some cases, the organism may suffer from a lack of potassium (hypokalemia). In cases of Ba-poisoning (not seen in welders), Ba damages the central and peripheral nervous system and the muscles including the heart [10].

Chromium(III) compounds (Cr_2O_3, $FeCr_2O_4$, $KCrF_4$)—are considered, according to general occupational experience, as having a very low toxicity.

Other metallic oxides—with toxic effects.

Zinc oxide (ZnO) (and also Copper oxide (CuO))—may cause metal (welding) fume fever (zinc oxide) or is suspected to cause metal fume fever (copper oxide). This results, after a latency period of some hours, with chills, fever, dyspnoe and often a metallic smell. The symptoms are self-limiting within a few days, but repeated metal fume fever attacks may result in bronchial hyperreactivity. The physiological background is a systemic inflammation which also includes the airways. There seems to be a disposition of particular persons, as not all welders exposed develop these symptoms. Typically, after repeated exposure, the symptoms decline ("Monday fever") [11].

2.2.3 Carcinogenic Substances

Chromium (VI) compounds—(chromates: Na_2CrO_4; K_2CrO_4) are carcinogenic to humans, resulting in cancer of the lungs and the airways. They are also an irritant to the airways.

Nickel oxides (NiO, NiO_2, Ni_2O_3) are carcinogenic to humans, resulting in cancer of the lungs and the upper airways.

Cadmium oxide (CdO)—is classified as carcinogenic. It acts as a strong irritant and results in possibly severe lung edema, often after slight acute symptoms and with a latency of several hours up to 2–3 days. After long term exposure, lung emphysema and also toxic kidney damage have been described.

Lead oxide (PbO)—is suspected of being carcinogenic. It may cause nausea, gastrointestinal symptoms, anemia, damage of the peripheral nerves and the kidneys. Intoxication have been seen in demolition work of constructions (i.e. by flame cutting) on constructions that had been coated with lead containing primers.

Beryllium oxide (BeO)—is classified as carcinogenic. Beryllium generally has a considerable toxic effect. The inhalation of fume and dust containing beryllium causes severe irritant effects in the upper respiratory tract and may result in severe lung disease (berylliosis, acute and chronic beryllium disease).

Cobalt oxide (CoO)—is classified as suspected carcinogenic substance. At higher exposures it is irritant to the airways. In combination with metal carbides, i.e. in cobalt based alloys, it may be a cause of lung fibrosis (hard metal disease), which can be seen in workers sintering and grinding hard metals (not seen in welders).

2.2.4 Radioactive Substances

Thorium dioxide (ThO_2) is a radioactive substance. The inhalation of fume and dust containing thorium dioxide leads to an internal radiation exposure.

Damages may occur in the shape of lung cancer, tumours of the bones and of other sites.

Under particular conditions, welders grinding or welding with Th-containing Wolframelectrodes may be categorized as "exposed to radiation" [12, 13]. Figure 2.2 on p. X shows a table summarizing the effects of the most important particulate hazardous substances on the human body.

References

1. Spiegel-Ciobanu, V. E. *Hazardous substances in welding and allied processes*. German Social Accident Insurance (DGUV), Information 209-017.
2. Spiegel-Ciobanu, V. E. *Nitrous gases in welding and allied processes, DGUV Information 209-048 BGI743E*.
3. Spiegel-Ciobanu, V. E., & Zschiesche, W. (2014). Best practice document on exposure to nitrogen oxides (NO/NO₂). *Welding in the World, 58,* 499–510.
4. Cosgrove, M. P., & Zschiesche, W. (2016). Arc welding of steels and pulmonary fibrosis. *Welding in the World, 60,* 191–199.
5. Kiesswetter, E., Schaeper, M., Buchta, M., et al. (2007). Longitudinal study on potential neurotoxic effects of aluminium: I. Assessment of exposure and neurobehavioural performance of Al welders in the train and truck construction industry over 4 years. *International Archives of Occupational and Environmental Health, 81,* 41–67.
6. Spiegel-Ciobanu, V. E., Brand, P., Lenz, K., et al. (2014). Characterisation of the biological effect of ultrafine particles in welding fumes after controlled exposure—Effect of the MIG welding of aluminium and the MIG brazing of zinc-coated materials. *Welding and Cutting, 13*(3), 171–176.

7. Spiegel-Ciobanu, V. E. (2006). Parkinson's disease and exposure to manganese during welding. *Welding and Cutting, 5*(2), 106–111.
8. McMillan, G., & Spiegel-Ciobanu, V. E. (2007). Manganism, Parkinson's disease and welders' occupational exposure to manganese—Part 1: Sources of manganese exposure and its role and function in human health and disease. *Welding and Cutting, 6,* 161–165.
9. McMillan, G., & Spiegel-Ciobanu, V. E. (2007). Manganism, Parkinson's disease and welders' occupational exposure in manganese—Part 2: Manganense as a neurotoxicological risk to welders. *Welding and Cutting, 6,* 220–229.
10. Zschiesche, W., Schaller, K.-H., & Weltle, D. (1992). Exposure to soluble barium compounds: An interventional study in arc welders. *International Archives of Occupational and Environmental Health, 64,* 13–23.
11. Markert, A., Baumann, R., Gerhards, B., et al. (2016). Single and combined exposure to zinc- and copper-containing welding fumes lead to asymptomatic systemic inflammation. *Journal of Occupational and Environmental Medicine, 58,* 127–132.
12. Ludwig, T., & Spiegel-Ciobanu, V. E. *Handling of thoriated tungsten electrodes during tungsten inert gas welding, DGUV Information 209-050 BGI 746E: (TIG).*
13. Costa, L. (2015). Welding with non-consumable thoriated tungsten electrodes. *Welding in the World, 59,* 145–150.

Chapter 3
Assignment of Hazardous Substances to Welding Processes and Materials

The studies described in clause 1.7 have yielded the following important results:

- the chemical composition of gaseous and particulate hazardous substances depends on the process and material used;
- hazardous substances never occur as a single component but always as a mixture of several components;
- depending on the process and material, one, two or even three components (gases and particles) may be predominant as far as their concentration and efficacy is concerned (e.g. the relevant limit values are the first to be exceeded).

Any predominant hazardous substance is called a **key component** (for a specific combination of process and material). A **main component** of the welding fume is a component of occupational medical importance, the percentage of which in the welding fume is not predominant, i.e., this component is no key component of the welding fume. Main components shall not be equated with key components.

In the following text the processes are divided into four main groups:

- welding;
- thermal cutting;
- thermal spraying;
- soldering and brazing.

The fume emission rates in this chapter are for illustrative purposes only and shall not be used for risk assessment purposes (see Chap. 5).

3.1 Welding

Welding always generates gaseous and particulate hazardous substances. The particulate substances have a particle size (aerodynamic diameter) of less than 1 μm, they are respirable and are normally called "welding fume". From the occupational health point of view, the respirable fraction is of special importance. This fraction, formerly

© International Institute of Welding 2020

V. E. Spiegel-Ciobanu et al., *Hazardous Substances in Welding and Allied Processes*, IIW Collection, https://doi.org/10.1007/978-3-030-36926-2_3

called fine dust, is usually measured with the sampling head for the inhalable fraction (formerly called total dust) during personal measurements carried out in welding. The reason is that at present it is still difficult to position the sampling head for the respirable fraction behind the welder's shield (lack of space).

Since welding only produces very fine particles, all of which are included in the "respirable fraction", measurements of "total dust" instead of "fine dust" are always on the safe side. The amount of hazardous substances generated in different welding processes is different. Today it is possible with systems like PGP-EA as in Germany to measure the respirable and the inhalable fractions simultaneously [1] (see clause 6.2).

Fume emission (mg/s) in welding is usually lower than fume and dust emission in cutting or spraying.

Studies on the emission of hazardous substances in welding have shown that approx. 95% of the welding fume is generated from the filler metal and only less than 5% from the parent metal.

3.1.1   Gas Welding

Gas welding of unalloyed and low-alloy steel mainly produces nitrous gases (nitrogen oxides). Here as for other oxy-fuel gas processes, e.g. flame heating and flame straightening, where an even larger amount of nitrogen oxides is generated, the key component is nitrogen dioxide.

The concentration of nitrogen dioxide in the air at the workplace increases with the length of the flame and hence with the size of the torch and the distance between tip and sheet.

The concentration of nitrogen dioxide becomes critical for operations in confined spaces without adequate ventilation measures. With a freely burning flame, it can reach 10 times the value produced by a flame with a length of 15 mm.

Measurements of emission rates in gas welding and heating revealed the following approximate values for nitrous gases:

Process	NO_x emission rates (mg/s)
Gas welding	0.8–40
Heating	up to 75

These values are high compared with emission rates at different arc welding processes, and the measured nitrous gases concentrations at the workplace will exceed the occupational limit values. It is necessary to control these concentrations at the workplace.

Problems with respect to generated airborne particles can only arise from the treatment of non-ferrous metals (e.g. lead, copper) and from coatings containing such metals.

3.1.2 Manual Metal Arc Welding with Covered Electrodes

Unalloyed, low-alloy steel (alloying components < 5%)
Compared with gas welding, this process generates high amounts of airborne particles.

Hazards caused by nitrous gases are unlikely.

In manual metal arc welding with unalloyed or low-alloy electrodes, the welding fume (total) shall be considered.

The chemical composition of the welding fume reflects the chemical composition of the core wire and of the covering. In this case, the main constituents of the welding fume are iron oxide (Fe_2O_3), silicon dioxide (SiO_2), potassium oxide (K_2O), manganese oxide (MnO), sodium oxide (Na_2O), titanium dioxide (TiO_2), aluminium oxide (Al_2O_3).

Dependent on the type of covering (acid, rutile, basic, cellulosic) these components occur in different proportions. The fume from basic covered electrodes also contains calcium oxide (CaO) and fluorides (F^-). Here, fluorides should be considered as another main component (Fig. 3.1).

Fume of electrodes with acid covering contain up to 10% of manganese oxide. Thus, **manganese oxide** may become an additional main component of the welding fume in this case.

Welding fume components	Type of coating			
	acid %	rutile %	basic %	cellulose %
Na_2O	2 - 4	2 - 4	2 - 4	2 - 4
Al_2O_3	1 - 2	1 - 2	1 - 2	1 - 2
SiO_2	30 - 40	30 - 40	~ 10	~ 10
K_2O	10 - 20	10 - 20	20 - 30	–
CaO	1 - 2	1 - 2	15 - 20	–
TiO_2	< 1	~ 5	~ 1	~ 1,5
MnO	~ 10	~ 7	~ 6	~ 5
Fe_2O_3	~ 40	20 - 30	20 - 30	70 - 80
F^-	–	–	12 - 16	–

Fig. 3.1 Analysis of welding fume generated by manual metal arc welding with unalloyed/low-alloy electrodes. *Source* [15], based on [16]

The following approximate emission values for welding fume are the result of many measurements carried out during manual metal arc welding with unalloyed/low-alloy electrodes:

Process	Welding fume emission rates (mg/s)
Manual metal arc welding	4–18

For special electrodes containing copper, copper oxide (CuO) may be an additional main component.

Chromium-nickel steel ($\leq$20% Cr and $\leq$30% Ni)

In addition to iron and substances from the coating (as above), high alloy covered electrodes contain up to 20% of chromium and up to 30% of nickel in the core wire.

During manual metal arc welding with high-alloy electrodes, welding fume is generated, the chemical composition of which may contain up to 16% of chromium compounds. Up to 90% of these chromium compounds are present as chromates (chromium(VI) compounds) which in most cases are classified as carcinogenic. Here, nickel oxide with 1% and seldom up to 3% is clearly under-represented.

For this process with the above materials, the key component in the welding fume is "*chromate*". The fume from basic covered electrodes contains much higher proportions of chromium(VI) than those from rutile electrodes.

Examination of biological material and epidemiological studies indicate that the strongest risk to a welder's health is produced by manual metal arc welding with high-alloy electrodes. Specific protective measures should be provided at the workplace, e.g. exhaust of welding fume at the point of origin. In addition, preventive occupational medical examinations shall be carried out.

Emission measurements made during manual metal arc welding with high-alloy electrodes gave the following approximate emission values for welding fume:

Process	Welding fume emission rates (mg/s)
Joint welding	2–16
Cladding	3–22

Nickel, nickel alloys (>30% Ni)

In manual metal arc welding with pure nickel or with nickel-base alloys, nickel oxide is the key component, even though the welding fume only contains a maximum of 5% of nickel oxide. Nickel oxides are classified as carcinogenic substances into category 1. Specific protective measures should therefore be provided at the workplace.

Apart from nickel oxide in the welding fume copper oxide may be generated—depending on the type of alloy (with copper components)—as another main component. When cladding with electrodes containing cobalt, attention should be paid to cobalt oxide (CoO).

Emission measurements made during manual metal arc welding with pure nickel and nickel-base alloys gave the following approximate welding fume emissions:

Process	Welding fume emission rates (mg/s)
Manual metal arc welding	approx. 7

3.1.3 Gas-Shielded Arc Welding, in General

In processes with active gas (MAGC, MAGM) primarily generation of large quantities of particulate hazardous substances (welding fume) can be expected. The amount of hazardous substances is of the same order as in manual metal arc welding with covered electrodes.

In contrast, processes using inert gas (MIG, TIG) exhibit a significantly lower fume generation.

Subject to the consumables and shielding gases used, gases and welding fume are generated, from which the key components are chosen. Figure 3.2 on p. XX gives some examples.

3.1.4 Gas-Shielded Metal Arc Welding (GMAW)

Emission rates may vary a lot depending on the shielding gas, the welding parameters and the type of consumable. From an industrial perspective, it is always very important to consider that Deposition rates (i.e. Kilograms of deposited metal per minute) are significantly affected by the same parameters, hence the exposure may be influenced.

In metal active gas welding with pure carbon dioxide (MAG) of unalloyed and low-alloy steel, carbon monoxide is a key component besides welding fume. By thermal decomposition of the carbon dioxide which is used as shielding gas, carbon monoxide is generated. Here, the welding fume is mainly composed of iron oxides.

Emission measurements made during MAG welding of unalloyed/low-alloy steel gave the following approximate emission values for welding fume and carbon monoxide:

Hazardous substance	Emission rates (mg/s)
Welding fume	2–12
Carbon monoxide (CO)	2–12.5

Pure carbon dioxide is not commonly used as a shielding gas to weld chromium-nickel steels.

In metal active gas welding with gas mixture (MAGM) of unalloyed and low-alloy steel, a gas mixture is used as shielding gas. If the gas mixture contains carbon dioxide, formation of a certain amount of CO is to be expected. Here, the welding fume is composed of iron oxides.

During welding of chromium-nickel steels, nickel oxide should be considered as possible key component. Although the welding fume contains up to 17% of chromium compounds and up to 5% of nickel oxide, the chromium compounds are almost exclusively composed of the trivalent form, which is not regarded as being carcinogenic.

Inert gasses are commonly used to weld non-ferrous metals. In metal inert gas welding (MIG) of aluminium-base materials, formation of ozone (from UV-radiation and from strongly reflecting materials) should be considered in addition to the welding fume (in the form of aluminium oxide). In many cases fume generation may be, however, lower than in MAG welding. Ozone concentrations are higher with aluminium-silicon-alloys than with pure aluminium and considerably higher than with aluminium-magnesium alloys.

When MIG welding nickel and nickel-base alloys, nickel oxide is the important key component in the welding fume. Due to the high nickel proportion in the filler metals the content of nickel oxide in the welding fume may reach values between 30 and 87% [2].

Emission measurements made during MIG welding of nickel and nickel-base alloys gave the following approximate emission values for welding fume and nickel oxide:

Hazardous substance	Emission rates (mg/s)
Welding fume	2–6
Nickel oxide	up to 5

Higher emission values of welding fume are generally to be expected for nickel base alloys containing copper than for nickel base alloys with other alloy elements such as Cr, Co, Mo. Here, copper oxide shall be considered as key component instead of nickel oxide. Protective measures shall be provided as for all other carcinogenic substances. Checking of ozone concentrations may also be necessary.

The electrode type is also influencing the fume generation amounts, as follows:

– Larger amounts of welding fume are generated when FCAW welding with flux-cored wire electrodes than when GMAW welding with solid wire electrodes, even if deposition rates may be significantly higher.
– The use of some self-shielded flux-cored wire electrodes has shown to be generating considerably higher welding fume emissions than the use of flux-cored wire electrodes under shielding gas.

However, when comparing flux-cored wires with metal-cored wires and solid wires, it is important to take wire diameter, shielding gas and welding parameters into account as they may significantly affect the emissions. As example, metal active gas welding of unalloyed and low-alloy steel gave the following emission values:

Filler metal	Welding fume emission rates (mg/s)
GMAW (with solid wire)	2–15
FCAW (with flux-cored wire with shielding gas)	5–55
MCAW (with metal-cored wire)	5–20
Self-shielded flux cored wire	2–100

In principle, the flux core of the wires contains components similar to those in the covering of a corresponding electrode.

Depending on the type of filler metal, different components of industrial hygiene relevance might be found in the welding fumes. ISO TR ISO/TR 13392:2014 "Health and safety in welding and allied processes—Arc welding fume components" reports an exhaustive list of those.

3.1.5 Tungsten Inert Gas Welding (TIG)

In tungsten inert gas welding (TIG), the formation of ozone is promoted by the reduced level of fume generation. Ozone values are especially high (but lower than for MIG) with pure aluminium and—even more so—with aluminium-silicon alloys. If pure nickel and nickel alloys are welded, nickel oxide may be the key component.

When using thoriated tungsten electrodes in TIG welding, especially when welding aluminium materials, a radiation exposure by inhalation of fume containing thorium dioxide can be expected. Limit values for persons not occupationally exposed to radiation during "work activities" are in general exceeded. Therefore, effective protection measures shall be provided at the workplace (e.g. use of non-thoriated tungsten electrodes). See also [3, 4].

3.1.6   Submerged Arc Welding (SAW)

With Submerged arc welding the arc is completely covered by a flux layer. There is no visible arc and the fume emission is very limited. Expressed in FER it is in general less than 1 mg/s.

3.1.7   Resistance Welding

In resistance welding with different materials, welding fume concentrations (metal oxides from spatter or evaporation of the material) are generated, which under normal operation and ventilation conditions, are generally below the reference levels.

Welding of oiled or greasy sheet steel should be avoided in practice, if possible. Thick layers of oil or grease lead to higher fume concentrations including proportions of organic substances.

When welding without spatter, about 30% more fume are generated from greased sheet than from non-greased sheet.

Compared to other resistance welding processes (e.g. spot welding), flash welding produces greater amounts of fume which usually necessitate exhaust at the machine.

3.1.8   Laser Welding Without Filler Metal

The use of lasers in welding and allied processes represents a relatively new and complex process. See also [5–8].

The high energy of the laser source causes evaporation from the parent metal (fused metal).

This leads to emissions of hazardous substances (welding fume), the chemical composition of which approximates that of the parent metal.

The amounts of hazardous substances formed during laser welding without filler metal are comparable to those formed during metal active gas welding. For example, laser welding of chromium-nickel steel produces emissions of 1.2–2 mg/s for total dust.

Emission measurements made when laser welding different metallic materials showed the following emission values for constant welding parameters (thickness of material = 1 mm, laser power = 2900 W, focal length = 200 mm, feed rate = 50 mm/s, process gas = Ar):

Material	Emissions of particulate hazardous substances [mg/s] (above processing side)
Unalloyed steel	1.5
Cr-Ni stainless steel	1.2
Galvanized steel	7
Titanium	0.9

The highest emissions of hazardous substances are observed for galvanised steel, where fume is essentially generated from the zinc coating.

Results show lower emission rates for gaseous hazardous substances:

Material	Emissions of gaseous hazardous substances (μg/s)		
	NO_x	CO	O_3
Unalloyed steel	200	56	53
Cr-Ni stainless steel	350	28	19
Galvanized steel	800	56	< detection limit

3.1.9 Laser Cladding

The use of lasers in welding and allied processes represents a relatively new and complex process.

In laser cladding, the filler metal can be added in the form of wire or powder. Mainly particulate hazardous substances (fume) are generated. If the filler metal is added in the form of powder, partially inhalable but non-respirable particulate substances are produced besides the fume. Total emissions of particulate substances during laser cladding are less than 5 mg/s.

Gaseous hazardous substances do not present a problem. The chemical composition of the welding fume is roughly similar to the chemical composition of the filler metal, elements with a low boiling temperature being overrepresented in the fume.

Apart from the key component (corresponding to the basic alloying element) oxides of other alloying elements (more than 10%) may also reach critical values after different periods of operation. These are the main components.

Laser cladding with cobalt based alloys produces welding fume and dust in which cobalt oxide is the key components.

For nickel based alloys which also include more than 10% cobalt, nickel or cobalt oxide may be the key component in the welding fume—depending on the respective percentage in the welding fume. The welding fume also contains aluminium oxide.

For laser cladding of iron-based alloys containing a high level of chromium, welding fume (iron oxide) shall be considered. Total chromium in the welding fume is mainly present in the metallic or trivalent oxide form. Measured chromium(VI) compound levels are very low ($\leq 5\%$ of total chromium).

Copper oxide should be considered as key component for complex aluminium bronze, due to the high content of copper (approx. 75% Cu). Besides, aluminium oxide is a main component.

3.1.10 Laser Welding with Nd:YAG-Laser

The use of lasers in welding and allied processes represents a relatively new and complex process.

Emissions of hazardous substances [emission rate, (mg/s)] are generally lower at optimum (welding) parameters when using solid state lasers (Nd:YAG-Laser) than when using CO_2 lasers, as, at present, the obtainable welding speeds are lower for the Nd:YAG laser than for the CO_2 laser [8, 9].

The absorbed intensity (power density) in the interaction zone is decisive for the amount of particulate substances. With an increase in intensity, the melting temperature and thus the evaporation rate increases.

Emission measurements made during welding of chromium nickel steel and galvanised steel as a function of the absorbed beam intensity gave the following welding fume emission rates:

Material	Welding fume emission rates (mg/s)
Chromium-nickel steel (s = 3 mm, v_s = 600 mm/min)	~1.5
Galvanised steel (s = 1 mm, v_s = 400 mm/min)	~2.7

The beam intensity varies between 3.18×10^5 and 6.67×10^5 W/cm^2.

3.1.11 Assignment of Key Components to Welding Processes (Example)

Figure 3.2 gives an example used in Germany for the assignment of key components to processes and materials used in welding.

Figure 3.3 gives an example used in Germany for the assignment of key components to processes and materials used in Laser welding.

Process	Filler metal	Welding fume/key component(s)
Gas welding	unalloyed, low-alloy steel (alloy components < 5%)	nitrogen monoxide nitrogen dioxide
Manual metal arc welding	unalloyed, low-alloy steel (alloy components < 5%)	welding fume[1] manganese oxide[4] (chromium(VI) compounds)[5]
	chromium-nickel steel (≤ 20% Cr and ≤ 30% Ni)	chromium(VI) compounds or manganese oxide[4]
	nickel, nickel alloys (> 30% Ni)	nickel oxide or copper oxide[2]
Metal active gas welding with carbon dioxide (MAGC)	unalloyed, low-alloy steel (alloy components < 5%)	welding fume[1] carbon monoxide
Metal active gas welding with gas mixture (MAGM)	unalloyed, low-alloy steel (alloy components < 5%) solid wire	welding fume[1] manganese oxide
	unalloyed, low-alloy steel (alloy components < 5%) flux cored wire	welding fume[1] manganese oxide (chromium(VI) compounds)[5]
	chromium-nickel steel solid wire (≤ 20% Cr and ≤ 30% Ni)	nickel oxide or manganese oxide[4] (chromium(VI) compounds)
	chromium-nickel steel flux-cored wire (≤ 20% Cr and ≤ 30% Ni)	chromium(VI) compounds or manganese oxide[4]
Metal inert gas welding (MIG)	nickel, nickel alloys (> 30% Ni)	nickel oxide or copper oxide[2], ozone
	pure aluminium aluminium-silicon-alloys	ozone welding fume (mainly Al)[1]
	other aluminium alloys (e.g. Al-Mg)[3]	welding fume (mainly Al)[1] ozone
Tungsten inert gas welding (TIG)	unalloyed, low-alloy steel (alloy components < 5%)	welding fume[1] ozone
	chromium-nickel steel (≤ 20% Cr and ≤ 30% Ni)	welding fume[1] ozone
	nickel, nickel alloys (> 30% Ni)	ozone welding fume[1]
	pure aluminium aluminium-silicon-alloys	welding fume (mainly Al)[1] ozone
	other aluminium alloys (e.g. Al-Mg)[3]	welding fume (mainly Al)[1] ozone

[1] Limit value for the respirable fraction of the dust
[2] Limit value for copper fume, depending on type of alloy, with/without copper
[3] Aluminium materials (pure aluminium, aluminium alloys) limit value for aluminium oxide fume
[4] When the manganese proportion in the alloy or the sum of proportions (alloy and cover/filling) is ≥ 5%
[5] For special filler metal with about 2,2 % chromium in the alloy

Fig. 3.2 Assignment of key components to processes and materials used in welding.
Source Spiegel-Ciobanu

Process	Parent metal	Welding fume/ key component(s)
Laser beam welding[1]	unalloyed, low-alloy steel	welding fume[2]
	chromium-nickel steel ($\leq$ 20% Cr and $\leq$ 30% Ni)	nickel oxide
	galvanised steel	zinc oxide
Process	**Filler metal**	**Key component(s)s**
Laser beam overlaying	cobalt base alloys (> 60% Co, > 20% Cr)	cobalt oxide
	nickel base alloys (> 60% Ni)	nickel oxide
	iron base alloys (> 40% Cr, > 60% Fe)	welding fume[2]
	complex aluminium bronzes ($\approx$ 75% Cu)	copper oxide[3]

[1]   here without filler metal
[2]   the limit value for the respirable and inhalable fraction should be taken
[3]   limit for copper fume

Fig. 3.3 Assignment of key components to processes and materials used in laser welding. *Source* Spiegel-Ciobanu

3.1.12 Hybrid Welding

Hybrid welding (=combination of two individual processes) gains increasing importance for welding fabrication. The best known hybrid processes are:

- Laser—GMAW,
- laser—GTAW,
- plasma—GMAW,
- plasma—GTAW
- laser—plasma welding.

These processes are based on a high power density energy source, interacting with an electric arc and a possible feeding unit. Depending on the specific features of the process the hazardous substances generated (mg/s) may be significantly different than for simple MIG or TIG welding.

For a correct design of the ventilation system, examinations for the determination of emission rates specific to process/material are advisable.

In the specific case of **Laser beam plasma welding of aluminium materials**, a significant increase in welding speed is achieved in respect to Laser welding. Here, higher emissions of hazardous substances have to be expected than for simple laser beam welding. The key components ozone and aluminium containing welding fume—which occur at the same time—shall be considered, even if other components may be found as a result of the material welded. An effective extraction may be installed directly in the generation zone of the above hazardous substances.

3.2   Thermal Cutting

This process group includes oxygen cutting, plasma cutting and laser cutting (Fig. 3.4 on p. X). The composition of the parent metal determines the chemical composition of the particulate substances (fume), the diameter of which is larger than in fume from welding but which are nevertheless respirable.

3.2.1   Flame Cutting (Unalloyed and Low-Alloy Steel)

This process produces high fume emissions as a function of different parameters

- sheet thickness
- fuel gas
- cutting gas pressure
- cutting speed.

In addition to welding fume—which is important in this process—generation of nitrous gases shall be considered, i.e. nitrogen dioxide shall be considered as key component besides welding fume.

In this process, the welding fume emission rate may vary from 10 to 50 mg/s [10].

3.2.2   Plasma Cutting

This process is usually accompanied by a significant emission of particulate substances.

The hazardous substances emitted mainly depend on the parent metal being cut (i.e. on its chemical composition), on the cutting parameters chosen, and on the kind of plasma gas used.

An increase in cutting speed (mm/min) leads to a reduction of the hazardous substances emitted (g/min).

For the processing of unalloyed and low-alloy steel welding fume (mainly iron oxides) and manganese oxides are relevant. In plasma cutting of chromium-nickel steel, however, nickel oxide is generated as the key component. In addition chromium (VI)-compounds are generated.

Nickel and nickel base alloys which are processed by plasma cutting produce high levels of nickel oxide in the welding fume.

For aluminium-base materials where the parent metals are strongly reflective (e.g. aluminium-silicon alloys) ozone may be a key component in addition to welding fume.

If no technical protective measures, such as down-draught exhaust, are provided, it can be assumed that the limit value for the respirable fraction and for the inhalable fraction will be exceeded at workers position at the machine, independent of the chemical composition of the materials If the materials contain more than 5% of chromium and nickel (high-alloy materials) this also applies to the former limit values for chromium(VI) compounds and nickel oxide, nickel oxide being the key component.

If compressed air or nitrogen is used as plasma gas, nitrogen dioxide should be considered as an additional key component.

3.2.3 Laser Beam Cutting, in General

Due to the complexity of processes and equipment, generation of hazardous substances in laser cutting is determined by many characteristics. See also [5, 11, 12].

In addition to the material and the parameters determined by the process, the laser source plays an important role in the generation and composition of hazardous substances.

Process	Parent metal	Welding fume/key components
Flame cutting	unalloyed, low-alloy steel (alloy components < 5 %)	welding fume[2] nitrogen dioxide
Plasma cutting[1] Laser cutting	unalloyed, low-alloy steel (alloy components < 5 %)	welding fume[2]
	chromium-nickel steel (≤ 20% Cr and ≤ 30% Ni)	nickel oxide
	nickel, nickel compounds (> 30% Ni)	nickel oxide
	aluminium-base materials[3]	welding fume[2] ozone

[1] When compressed air or nitrogen is used as plasma gas, nitrogen dioxide should also be considered as key component!
[2] Limit value for the A fraction of the dust
[3] Aluminium-base materials (pure aluminium, aluminium alloys)
     limit value for aluminium oxide fume

Fig. 3.4 Assignment of key components to process and material used in thermal cutting. *Source* Spiegel-Ciobanu

3.2.4 *Laser Cutting with CO_2-Laser*

Parameters which have a great influence on the amount of hazardous substances emitted are the thickness of the workpiece, the lens focal length, the cutting gas pressure, the laser beam power, and the cutting speed. Considering the latter, emissions are commonly referenced to the length of cut (mg/m).

Dust emissions (mg/m) increase with increasing workpiece thickness and/or with an increasing lens focal length and/or with an increasing cutting gas pressure and/or with an increasing laser beam power. With increasing cutting speed, dust emissions increase per time unit (mg/s) and decrease per cutting length (mg/m).

Laser cutting produces relatively large amounts of dust, however, the amount produced is less than in oxygen or plasma cutting.

The largest emissions of hazardous substances occur in laser cutting of chromium-nickel steel.

The cutting of galvanised steel results in higher emission rates than the cutting of unalloyed steel.

During cutting with CO_2-laser (power = 1 kW) the following welding fume emissions were obtained for equal workpiece thicknesses:

Material	Welding fume emission rates (mg/s)
Unalloyed steel	16–24
Chromium-nickel steel	14–35

With regard to the problem of hazardous substances, it is possible to differentiate between

- high-pressure laser cutting (with nitrogen) and
- laser cutting (with oxygen).

The use of nitrogen as cutting gas reduces the emissions of hazardous substances from chromium-nickel steel and galvanised steel by half compared to cutting with oxygen.

Welding fume emissions in high-pressure laser cutting and laser cutting (example): (Power = 1 kW, focal length of lens = 63.5 mm, thickness of workpiece = 1 mm)

Material	Welding fume emission rates (mg/s)	
	High-pressure laser cutting processing gas: Nitrogen	Laser flame cutting processing gas: Oxygen
Low-alloy steel	–	17
Chromium nickel steel	8	20
Galvanised steel	4.5	9

Without adequate exhaust systems, the limit values for the corresponding key components are exceeded in laser cutting, irrespective of the parent metal. See also [6].

3.2.5 Laser Cutting with Nd:YAG-Laser

Here again (as in clause 3.1.6) the total emissions of hazardous substances are lower for the use of solid state lasers (Nd:YAG lasers) than for the use of CO_2 lasers (Fig. 3.5), whereas the cutting speed which can presently be reached is lower for Nd:YAG lasers than for CO_2 lasers [5, 8, 9, 11].

During cutting with an Nd:YAG laser the following welding fume emissions were obtained for a material thickness of 1 mm:

Material	Welding fume emission rates (mg/s)	
	Processing gas	
	N_2	O_2
Chromium-nickel steel ($l = 2.98 \times 10^6$ W/cm^2)	(v_s = 850 mm/min) ~2	(v_s = 400 mm/min) ~2.7
AlMg 3 ($l = 1.89 \times 10^6$ W/cm^2)	(v_s = 200 mm/min) ~0.3	–

Here, welding fume emission (mg/m) increases with the increase of the material thickness. If nitrogen is used as processing gas (cutting gas), welding fume (mg/s) is considerably reduced.

The parameters having an essential influence on the amount of emitted hazardous substances are: absorbed intensity as a function of power density, processing gas pressure, cutting speed, processing efficiency degree, thickness of the workpiece.

The amounts of particulate hazardous substances generated during the use of Nd:YAG lasers are in the order of magnitude of those generated during metal inert gas welding (MIG), only for this example, with P= 400 W. Technical ventilation measures, especially effective extraction systems, may be necessary to avoid exceeding the limit values for the relevant key components.

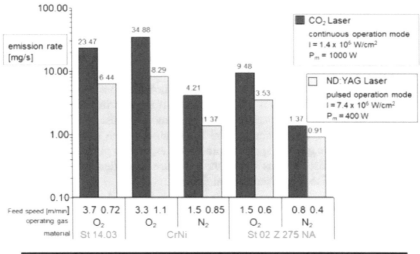

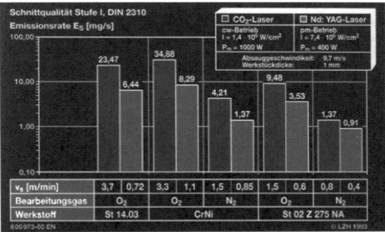

Material thickness for both techniques : 1 mm
Fume capture velocity for both techniques : 9.7 m/s

Fig. 3.5 Comparison of emissions generated during laser cutting with CO_2 and Nd:YAG lasers. *Source* Engel, p. 92, Fig. 44 [8, 9]. Legend: Schnittqualität Stufe I, DIN = Cutting Quality Level I, German Standard; Emissionsrate = emission rate; cw-Betrieb = continuos operation mode; pm-Betrieb = pulsed operation mode; Absauggeschwindigkeit = fume extraction velocity; Werkstückdicke = thickness of the plate; Bearbeitungsgas = operating gas; Werkstoff = material

3.3   Thermal Spraying

Thermal spraying produces large amounts of particulate hazardous substances, depending on the process used (Fig. 3.6), particularly as a function of the power density of the source

Thus, emissions of hazardous substances are significantly lower in flame spraying than in arc spraying, and plasma spraying produces the largest emissions of hazardous substances [13].

Furthermore, the hazardous substances generated depend on the material and are exclusively emitted from the spraying material; vice versa, the parent metal has no influence on the amount and the composition of the hazardous substances produced.

In all spraying processes, if not provided or if there is insufficient capture and separation of hazardous substances, welding fume and dust concentrations in the breathing zone exceed the occupational limit value for the respirable and for the inhalable fraction. In general, spraying processes (especially plasma spraying) should be carried out in closed booths so that the exposure of welders and other persons to fume, dusts and noise is reduced to a minimum.

3.3.1 Flame Spraying

Flame spraying using wires and powder as spraying materials generates gaseous and particulate substances. The chemical composition of the particulate substances in fume/dust corresponds to the composition of the spraying material. As in other oxy-fuel processes, generation of nitrous gases should be also taken into account.

During flame spraying with high-alloy spraying material (e.g. chromium $< 27\%$, Ni $< 22\%$) high levels of dust emissions include also high proportions of nickel oxide.

In this process, nickel oxide concentrations may considerably exceed 0.5 mg/m^3 and Chromium (VI) compounds may in addition be generated (it is assumed that a varied mixture of different chromium oxides is produced; this mixture, hardly soluble, may contain chromium (VI) compounds and is regarded as carcinogenic).

Nickel oxide is the key component when nickel and nickel alloys are used. Here again, 0.5 mg/m^3 is expected to be exceeded frequently.

At the same deposition rate, chromium-nickel alloys produce higher emissions than zinc or aluminium alloys.

3.3.2 Arc Spraying

Arc spraying produces large emissions of particulate substances. For comparable spraying parameters and at approximately the same deposition rate, the emission of hazardous substances from aluminium wire is higher than that from zinc, chromium, nickel and aluminium bronze wires, where the emissions of hazardous substances are comparable. During arc spraying with chromium-nickel or nickel-base spraying materials, nickel oxide shall be considered as the key component. Here is also the use of an effective exhaust system necessary.

The diameter of particles is usually smaller in arc spraying than in flame spraying, resulting in a larger respirable fraction.

3.3.3 Plasma Spraying

Plasma spraying produces higher emissions of hazardous substances than flame or arc spraying with the same spraying materials, due to the use of a much higher spraying rate.

Process	Spraying material	Welding fume/ key component(s)
Flame spraying	unalloyed, low-alloy steel (alloying components > 5 %)	respirable fraction, inhalable fraction[1] nitrogen dioxide
	chromium-nickel steel ($\leq$ 27% Cr and $\leq$ 22% Ni)	nickel oxide nitrogen dioxide
	nickel and nickel alloys (> 60% Ni)	nickel oxide nitrogen dioxide
	aluminium-base materials[3]	respirable fraction, inhalable fraction[1] nitrogen dioxide
	lead alloys	lead oxide nitrogen dioxide
	copper and copper alloys	copper oxide[2] nitrogen dioxide
	other non-ferrous metals and alloys	respirable fraction, inhalable fraction[1] nitrogen dioxide
Arc spraying	unalloyed, low-alloy steel (alloying components < 5 %)	respirable fraction, Inhalable fraction[1]
	chromium-nickel steel ($\leq$ 27% Cr and $\leq$ 22% Ni)	nickel oxide
	nickel and nickel alloys (> 60% Ni)	nickel oxide
	aluminium-base materials[3]	respirable fraction, Inhalable fraction[1]
	copper and copper alloys	copper oxide[2]
	other non-ferrous metals and alloys	respirable fraction, Inhalable fraction[1]
Plasma spraying	copper aluminium and copper tin alloys	copper oxide[2]
	chromium nickel steel ($\leq$ 27 % Cr and $\leq$ 22 % Ni)	nickel oxide ozone
	nickel and nickel alloys (> 60 % Ni)	nickel oxide
	cobalt base alloys (> 50% Co)	cobalt oxide

[1] Limit value for respirable dust/welding fume and inhalable dust
[2] Limit value for copper fume
[3] Aluminium-base materials (pure aluminium, aluminium alloys) limit value for aluminium oxide fume

Fig. 3.6 Assignment of key components to processes and materials used in thermal spraying. *Source* Spiegel-Ciobanu

Most of the plasma spraying processes are therefore carried out in enclosed systems (encapsulated systems). Nevertheless, there may still be an health risk for the operator for the few manual spraying processes, if the high hazardous substance concentrations are not exhausted at source.

Practice shows that occupational limit values may be substantially exceeded during plasma spraying with materials having higher proportions of critical materials (chromium, nickel, cobalt, etc.) if no effective exhaust system is in operation.

3.4 Soldering and Brazing

Here again, the emission of hazardous substances is related to the process and the material used. The amount and the chemical composition of hazardous substances generated (soldering or brazing fume) depend on the materials used (solder and brazing alloys, flux, binder) (Figs. 3.10 and 3.11) and on the process parameters (brazing or soldering temperature, Fig. 3.7, soldering and holding time).

Considering that the base material is not melted, its composition is not expected to influence the composition of the fumes.

Soldering and brazing are principally classified according to process temperature.

3.4.1 Soldering (T < 450 °C)

The generation of hazardous substances first of all depends on the soldering temperature. See here Fig. 3.7.

The filler metals (solder alloy) used are mostly based on tin, with addition of other elements; for some applications (e.g. high reliability products) lead containing alloys may be used.

The fluxes in soldering are mainly based on colophony, other organic resins or abietic acid. Colophony is a sensitizing agent to the skin. Thermal decomposition products of the fluxes may contain aldehydes such as formaldehyde and others, which are strong irritants to the airways. There have been clusters of acute asthma bouts described in solder workers.

Soldering temperature [°C]	Emission			
	Total fume [mg/g flux]	Colophony [mg/g solder]	Aldehyde [mg/g solder]	Tin [µg/g solder]
250	40	1	2×10^{-3}	8
450	102	4,2	$12,5 \times 10^{-3}$	30

Fig. 3.7 Emission of hazardous substances as a function of the soldering temperature. *Source* [15], modified from [10]

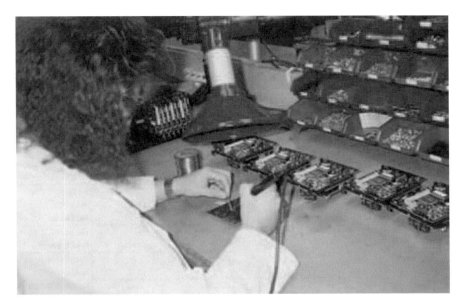

Fig. 3.8 Soldering workplace with local extraction. *Source* Courtesy of the German Liability Accident Insurance Institution for the Energy, Textile, Electrical and Media Products Sectors (BG ETEM)

In addition, hydrazine, lead, hydrogen chloride and bromide or tin compounds may be found, according to the solder and flux used. This is mainly the case for soldering operations in installations (Fig. 3.8).

Due to the solders and brazing alloys used during soldering and brazing, a variety of hazardous substances may be generated. The following hazardous substances were found in soldering and brazing fumes, among others: aldehydes (especially formaldehyde, acetaldehyde, acrylic aldehyde), antimony oxide, inorganic and organic tin compounds, lead oxide, boron oxide, boron trifluoride, cadmium oxide, chlorides/hydrogen chloride, fluorides, hydrogen fluoride, hydrazine, copper oxide, colophony, phosphorous pentoxide, silver oxide, zinc oxide.

Based on the type of flux used (see Fig. 3.9), and on the type of solder used the hazardous substances generated during soldering, which have to be taken into account in the framework of the hazard determination and evaluation, are listed in Fig. 3.10.

Group	Flux
1	natural resins (colophony) or modified natural resins with or without addition of organic or halogen containing activators
2	organic acids (e.g. citric, oleic, stearic, benzoic acid), amines, diamines, urea and organic halogen compounds
3	zinc and other metal chlorides, ammonium chloride (in aqueous solution or in organic preparations)

Fig. 3.9 Classification of fluxes into groups. *Source* [15], based on information from the German Liability Accident Insurance Institution for the Energy, Textile, Electrical and Media Products Sectors (BG BG ETEM)

Soldering (temperature < 450∞C)			
Solders		Fluxes (flux basis)	Key components
Field of application	Type of solder		
Heavy metals	antimony-containing, low-antimony, antimony-free lead-tin and tin-lead solders	Group 1	Respirable dust aldehyde[1] lead oxide
		Group 2	Respirable dust lead oxide[1]
		Group 3	Respirable dust lead oxide[1]
	tin-lead solders with copper, silver or phosphorus additions	Group 1	Respirable dust aldehyde[1] lead oxide
		Group 2	Respirable dust lead oxide[1]
		Group 3	Respirable dust lead oxide[1]
	tin solders with silver, copper, bismuth, indium, antimony and zinc additions	Group 1	Respirable dust aldehyde[1]
		Group 2	Respirable dust
		Group 3	Respirable dust
	cadmium solders with zinc, tin, silver and lead additions	Group 1	Respirable dust aldehyde[1] cadmium oxide
		Group 2	Respirable dust cadmium oxide
		Group 3	Respirable dust cadmium oxide
Light metals	solders on the basis of: - tin-zinc - zinc-cadmium - zinc-aluminium - lead-tin-silver	organic compounds, e.g. amines, organic halogen compounds	Respirable dust cadmium oxide
		chlorides, fluorides, e.g. zinc chloride	Respirable dust cadmium oxide chlorides fluorides
[1] Except for workplaces, where electric and electronic component groups or their individual components are soldered or at the repair workplaces in these areas			

Fig. 3.10 Key components during soldering. *Source* [15], based on information from the German Liability Accident Insurance Institution for the Energy, Textile, Electrical and Media Products Sectors (BG ETEM)

3.4.2 Brazing (T > 450 °C)

For brazing, brazing alloys may contain aluminium, copper, zinc, nickel, tin, silver and cadmium additives. The fluxes used contain mixtures of boric acids, single and complex fluorides and borax.

Depending on the brazing alloys and flux, brazing may produce hazardous substances such as cadmium oxide, copper oxide, zinc oxide, silver oxide, fluorides, boron oxides, etc.

From the occupational health point of view, cadmium compounds and fluorides in brazing fume are especially important.

During **brazing with alloys containing cadmium** under unfavourable hygienic conditions toxic life threatening events, resulting in lung edema have been described.

Brazing (Temp. ≥ 450°C)			
Brazing alloys for brazing		**Flux (flux basis)**	**Key component**
Field of application	**Type of brazing alloy**		
Heavy metals	silver containing brazing alloys, cadmium free	boron compounds with additions of single and complex fluorides, phosphates and silicates	Respirable dust chlorides fluorides silver oxide
	silver containing brazing alloys, cadmium-containing		Respirable dust chlorides fluorides silver oxide cadmium oxide
	phosphorous brazing alloys		Respirable dust chlorides fluorides
	zinc and zinc containing brazing alloys		Respirable dust chlorides fluorides zinc oxide
	copper and copper based brazing alloys[1]		Respirable dust chlorides fluorides copper oxide
	nickel base brazing alloys[1]		Respirable dust nickel oxide
	palladium containing brazing alloys[1]		Respirable dust
	gold containing brazing alloys[1]		Respirable dust
Light metals	aluminium base brazing alloys	chlorides and fluorides	Respirable dust chlorides fluorides
[1] In general, these brazing alloys are used in shielding gas ovens or in vacuum ovens without fluxes			

Fig. 3.11 Key components during brazing. *Source* [15], based on information from the German Liability Accident Insurance Institution for the Energy, Textile, Electrical and Media Products Sectors (BG ETEM)

Based on the type of flux used (see Fig. 3.9), the hazardous substances generated during brazing which have to be taken into account in the framework of the hazard determination and evaluation are listed in Fig. 3.11.

3.4.3 MIG Brazing, Laser Brazing, Plasma Brazing (T > 900 °C)

For these processes mostly copper base alloys in the form of wire are used as filler metal with a melting temperature lower than that of the parent metal, e.g. CuSi3 (Si 3%, Mn 1%, rest Cu), AlBz 8 (Al 8.2%, rest Cu). Hence, high amounts of copper oxide are generated from the filler metal.

As previously stated, the base material composition does not affect the exposure; however, during processing of galvanized steels, the fume contains high proportions of zinc oxide generated by vaporization and oxidization of the coating.

The highest emission rates must be anticipated for MIG brazing, while the lowest emission rates are generated during plasma and Laser brazing.

As examples:

– during MIG brazing with CuSi 3 (wire diameter of 1 mm) and a zinc layer of 45 μm, about 4.7 mg/s are generated;
– during MIG brazing with a wire diameter of 1 mm, the fume generation is about 2.4 mg/s.

During plasma brazing and laser brazing with the same filler metal, the welding fume emission is much lower than during MIG brazing [14].

References

1. Möhlmann, C., Aitken, R. J., Kenny, L. C., Görner, P., VuDuc, T., & Zimbelli, G. (2003, October). Größenselektive personenbezogene Staubprobenahme: Verwendung offenporiger Schäume. *Gefahrstoffe – Reinhaltung der Luft, 63*(10), 413–416.
2. Dilthey, U., & Holzinger, K. (1996 und 1998). *Untersuchungen zur Schadstoffentstehung beim MIG-Schweißen von Nickel- und Nickelbasislegierungen, Abschlussbericht.* Aachen: ISF-Aachen.
3. Ludwig, T., & Spiegel-Ciobanu, V. E. *Handling of thoriated tungsten electrodes during tungsten inert gas welding, DGUV Information 209-050 BGI 746E: (TIG).*
4. Costa, L. (2015). Welding with non-consumable thoriated tungsten electrodes. *Welding in the World, 59,* 145–150.
5. Zschiesche, W., Haferkamp, H., Seebaum, D, Goede, M., & Lehnert, G. (1995). Untersuchungen zur Gefahrstoffemission beim Laserschweißen und Laserschneiden verschiedener Stähle unter arbeitsmedizinischen undwelthygienischen Gesichtspunkten. In R. Schiele, B. Beyer, & A. Petrovic (Eds.), *Proceedings of the Annual Congress of the German Society for Occupational and Environmental Medicine,* Rindt Druck, Fulda, pp. 359–369.
6. Wittbecker, J.-G. *Gefahrstoffermittlung bei der CO$_2$-Laserstrahlbearbeitung. Reihe 2: Fertigungstechnik* Nr. 298. Düsseldorf: VDI-Verlag.

7. Klein, R. M., Dahmen, M., Peutz, H., et al. (1998). Workplace exposure during laser machining. *Journal of Laser Application, 10,* 99–105.

8. Haferkamp, H., Püster, T., & Engel, K. (1993, March 29–30). Investigations of hazardous emissions during laser surface treatment. In *Proceedings: Second Eureka Industrial Safety Forum'93, Coventry* (S. 89–104).

9. Haferkamp, H., Engel, K., Goede, M., Schmidt, H., & Bach, F.-W. (1994). Formation of process products during laser beam processing with Nd:YAG-lasers. In *Proceedings ICALEO'94, 13th International Congress on Applications of Lasers and Electro-Optics*, 17–20 October 1994.

10. Kraume, G., & Zober, A. (1989). *Arbeitssicherheit und Gesundheitsschutz in der Schweißtechnik* (Band 105). Düsseldorf: DVS-Verlag.

11. Bach, F. W., Haferkamp, H., Vinke, T., & Wittbecker, J. S. (1991). *Ermittlung der Schadstoffemissionen beim thermischen Trennen nach dem Laserprinzip*, Bundesanstalt für Arbeitsschutz und Arbeitsmedizin, Fb 615 (2nd edn.). Bremerhaven: Wirtschaftsverlag NW.

12. Zschiesche, W., Haferkamp, H., Engel, K., et al. (1994). Occupational and environmental medical aspects in laser beam cutting of selected thermoplastic polymers. *Arbeitsmed Sozialmed. Umweltmed, 29,* 512–516 (in German; English abstract).

13. Lauterbach, R. *Umweltbelastungen beim atmosphärischen Plasmaspritzen. Reihe Werkstofftechnik.* BIA-Report 6/94. Verlag Shaker, "Grenzwerteliste 1994".

14. Püster, T., & Walter, J. *Vergleichsuntersuchungen zur Bestimmung von Emissionsprodukten beim Löten mittels MIG, Plasmaquelle und Laserstrahl.* Abschlussbericht Juni 2000, LHZ Hannover.

15. Spiegel-Ciobanu, V. E. *Hazardous substances in welding and allied processes.* German Social Accident Insurance (DGUV), Information 209-017.

16. Eichhorn, F., Trösken, F., & Oldenburg, Th. (1981). *Untersuchung der Entstehung gesundheitsgefährdender Schweißrauche beim Lichtbogenhandschweißen und Schutzgasschweißen.* Düsseldorf: DVS-Verlag.

Chapter 4
Ultrafine Particles (UFP)

4.1 Findings from Studies on Toxicity of UFP in General

Epidemiological studies and animal tests with ultrafine particles non specifically connected with welding are in principle inhalation tests with rats or other rodents exposed to diesel soot, to technical soot and other particles to evaluate the effect of the size of particles independent of the chemical composition. They show that, depending on the particle surface in contact with the lung, inflammatory, fibrotic (hardening of the lung tissue) or carcinogenic effects may occur.

Studies show that particle size, structure and shape of the ultrafine particles contained in the welding fume can influence potential effects on the human organism. In addition, the chemical composition of the particles is of great importance [1, 2].

The particle concentration (particle/cm^3), the specific particle surface as well as the chemical composition have influence on the probability of deposits in the lung, the cleaning behaviour of the lung and the resulting staying time, thus determining the damage to the lung.

The largest damaging potential is assigned to particles having a size between 20 and 50 nm [1, 2].

4.2 Findings from Research on the Characteristics of UFP

Physico chemical processes such as vaporization, condensation, oxidation, pyrolysis caused by high temperatures during welding processes generate ultrafine particles.

The processes applied and their parameters (current, voltage, electrode diameter, type of welding) the filler and parent metal used with their chemical composition, the shielding gases or the flux material as well as the surface quality of the materials play an important role. The nature and amount of particles generated are highly

© International Institute of Welding 2020
V. E. Spiegel-Ciobanu et al., *Hazardous Substances in Welding and Allied Processes*, IIW Collection, https://doi.org/10.1007/978-3-030-36926-2_4

influenced by these factors. Furthermore, workplace specific factors such as ventilation, head/body position, duration of welding, shape/construction of the workpiece determines the welding fume concentration in the welders' breathing zone [3–5].

4.3   Results from Studies on Welding Processes

The ISF of RWTH Aachen in colaboration with ITEM Hannover has examined welding, cutting and soldering processes with unalloyed, low-alloy and high-alloy materials as well as aluminium-based and copper-based alloys concerning their emission rates, chemical composition, diffusion equivalent diameter of the agglomerates and the geometrical diameter of primary particles and with respect to "biologically active surface" of particles. The results are summarized in Figs. 4.1, 4.2, 4.3, 4.4, 4.5, 4.6 and 4.7 [5, 6] and are presented as follows.

With time, increasing sizes of particles and agglomerates occur so that an "ageing process" of the welding fume can be expected during measurements of welding fume concentrations at the workplace.

Mass and number size distributions, agglomerate number emission rates (number of agglomerates per second), medium primary particle number per agglomerate are subjected to changes by ageing.

Manual metal arc welding

Pr. #	Parent metal	Electrode	Gas	Mass-emission rate [mg/s]	MMAD [μm]	GSD (average)	Number median of mobility distribution	GSD (average)	Agglomerate emission rate (1/sec)	Mean primary particle number per agglomerate	calculated primary particle emission rate	Median value of primary particle size distribution [nm]	GSD (average)
3	1.0038	Rutile basic, low alloy.	no gas	5,8	0,37	3,0	129	1,8	4,2E+11	91	3,8E+13	7	2,1
4	1.0038	Rutile basic, low alloy	no gas	3,9	0,35	1,9	126	1,7	4,44E+11	89	4,0E+13	9	2,3
5	1.0038	basic, low alloy	no gas	7,6	0,45	1,5	135	1,6	5,2E+11	24	1,2E+13	15	2,2
6	1.0038	Cellulose	no gas	4,4	0,34	4,2	117	1,7	7,76E+11	76	5,9E+13	6	1,9
7	1,4301 basic, high alloy			5,5	0,44	1,6	142	1,7	3,57E+11	37	1,3E+13	13	2,2

Metal active gas welding / build-up welding

Pr. #	Parent metal	Electrode	Gas	Mass-emission rate [mg/s]	MMAD [μm]	GSD (average)	Number median of mobility distribution	GSD (average)	Agglomerate emission rate (1/sec)	Mean primary particle number per agglomerate	calculated primary particle emission rate	Median value of primary particle size distribution [nm]	GSD (average)
8	1.0038	Solid wire, low alloy	100% CO_2	9,6	0,37	1,7	156	1,6	8,59E+11	343	3,0E+14	15	2,1
9	1.0038	rutile cored wire, low alloy.	100% CO_2	12,9	0,40	1,5	142	1,6	6,88E+11	159	1,1E+14	13	2,1
10	1.0038	solid wire, low alloy	82% Ar, 18% CO_2	2,5	0,39	1,9	129	1,8	4,83E+11				
Primed 56 Sheet		solid wire, low alloy	82% Ar, 18% CO_2	2,9	0,21	2,3	152	1,8	3,25E+11	138	4,5E+13	11	1,6
11	1.0038	metal powder cored wire, low alloy	82% Ar, 18% CO_2	12,7	0,30	1,6	152	1,6	1,18E+12	330	3,9E+14	13	2,0
12	1.0038	rutile cored wire, low alloy.	82% Ar, 18% CO_2	16,3	0,37	1,4	149	1,5	1,3E+12	37	4,8E+13	33	1,6
13	1.4541	Solid wire, high alloy	90% Ar, 10% CO_2	3,7	0,19	1,8	149	1,7	5,09E+11	93	4,7E+13	8	1,9
14	1.4541	Solid wire, high alloy	97,5% Ar, 2,5% CO_2	2,9	0,13	1,5	129	1,6	7,05E+11	353	2,5E+14	11	1,7
15	1.4301	Solid wire, high alloy	90% Ar, 10% CO_2	2,5	0,45	3,3	120	1,6	1,24E+12				
16	1.4301	Solid wire, high alloy	97,5% Ar, 2,5% CO_2	3,2	0,20	2,1	138	1,6	1,1E+12	334	3,7E+14	13	1,6
17	1.4541	Cored wire, high alloy	90% Ar, 10% CO_2	3,1	0,27	1,9	129	1,6	6,47E+11				
18	1.4541	Cored wire, high alloy	97,5% Ar, 2,5% CO_2	4	0,27	2,0	126	1,6	5,51E+11	69	3,8E+13	23	1,7
19	1.0038	Metal powder cored wire, high alloy.	kein Gas	30,4	0,35	1,6	144	1,9	1,69E+12	185	2,1E+14	22	1,7
20	1.0038	Cored wire, low alloy.	82% Ar, 18% CO_2	24,8	0,39	3,0	152	1,6	1,13E+12				

Fig. 4.1 Characteristics of UFP in manual arc welding and metal active gas welding. *Source* [6, 7]

Metal inert gas welding

Pr. #	Parent metal	Wire	Gas	Mass emission rate [mg/s]	MMAD [µm]	GSD (average)	Number median of mobility distribution	GSD (average)	Agglomerate emission rate (1/sec)	Mean primary particle number per agglomerate	calculated primary particle emission rate	Median value of primary particle size distribution [nm]	GSD (average)
35 Al 99.5	S-Al 99.5	100% Ar		5,27	0,32	2,5	181	1,7	3,44E+11	264	9,1E+13	9	1,7
36 Al 99.6	S-Al 99.5	50% Ar, 50% He		6,9	0,36	2,0	194	1,7	2,85E+11	809	2,3E+14	7	1,6
37 AlSi1Mg	AlSi 5	100% Ar		9,28	0,37	1,9	190	1,7	3,67E+11	1535	5,6E+14	16	1,7
38 AlSi1Mg	AlSi 5	50% Ar, 50% He		7,54	0,42	1,8	164	1,7	2,74E+11	86	2,4E+13	9	1,9
39 AlSi1Mg	AlSi 12	100% Ar		3,16	0,33	1,9	154	1,7	3,35E+11	480	1,6E+14	14	1,6
40 AlSi1Mg	AlSi 12	50% Ar, 50% He		3,03	0,38	1,6	92	1,8	2,24E+11	295	6,6E+13	9	1,7
41 AlMg3	Al Mg 3	100% Ar		23,82	0,38	1,8	208	1,6	5,24E+11	1393	7,3E+14	14	1,7
42 AlMg3	Al Mg 3	50% Ar, 50% He		31,96	0,40	1,7	199	1,7	4,17E+11	189	7,9E+13	13	1,8
43 AlMg3	Al Mg 5	100% Ar		11,42	0,34	2,1	178	1,6	2,6E+11	1850	4,8E+14	16	1,7
44 AlMg3	Al Mg 5	50% Ar, 50% He		6,62	0,33	2,0	160	1,7	2,44E+11	973	2,4E+14	15	1,8
45 Nickel	Alloy 617	100% Ar		1,45			60	1,7	3,93E+11	116	4,5E+13	12	1,5

Metal inert gas soldering

Pr. #	Parent metal	Wire	Gas	Mass emission rate [mg/s]	MMAD [µm]	GSD (average)	Number median of mobility distribution	GSD (average)	Agglomerate emission rate (1/sec)	Mean primary particle number per agglomerate	calculated primary particle emission rate	Median value of primary particle size distribution [nm]	GSD (average)
47 DX53	AlBz25Ni2	97,5% Ar, 2,5%CO2		2,6	0,32	2,0	149	1,7	3,17E+11	571	1,8E+14	19	1,5
48 DX53	AlBz25Ni2	99,5% Ar, 0,5%O2		2,6	0,22	2,9	138	1,7	4,31E+11	499	2,2E+14	9	1,5
49 DX53	CuSi3	97,5% Ar, 2,5%CO2		2,3	0,17	2,0	129	1,7	2,91E+11	180	5,2E+13	11	1,6
50 DX53	CuSi3	99,5% Ar, 0,5%O2		1,9	0,16	1,9	135	1,7	3,84E+11	308	1,2E+14	13	1,5
51 DX53	CuAl8	97,5% Ar, 2,5%CO2		3,9	0,33	1,6	164	1,7	3,29E+11	735	2,4E+14	10	1,6
52 DX53	CuAl8	99,5% Ar, 0,5%O2		4,9	0,29	1,8	152	1,6	3,67E+11	149	5,5E+13	8	1,6

Fig. 4.2 Characteristics of UFP in metal inert gas welding and metal inert gas soldering. *Source* [6, 7]

Tungsten inert gas welding

Pr. #	Parent metal	Wire	Gas	Mass emission rate [mg/s]	MMAD [µm]	GSD (average)	Number median of mobility distribution	GSD (average)	Agglomerate emission rate (1/sec)	Mean primary particle number per agglomerate	calculated primary particle emission rate	Median value of primary particle size distribution [nm]	GSD (average)
64 1.4541	High alloy rod	100% Ar	0,0144			23	1,7	3,69E+11	17	6,2E+12	22	1,4
65 2.0090	copper rod	100% Ar	0,0142			24	1,6	6,14E+11				
66 AlMg3Mn	AlMg-alloy	70% Ar, 30% He	0,0276			32	1,6	4,04E+11	31	1,2E+13	15	2,0
67 AlSi1MgMn	AlSi-alloy	70% Ar, 30% He	0,0377			63	1,9	4,14E+11	120	5,0E+13	10	1,5

Laser beamweleding without filler metal

Pr. #	Parent metal	Wire	Gas	Mass emission rate [mg/s]	MMAD [µm]	GSD (average)	Number median of mobility distribution	GSD (average)	Agglomerate emission rate (1/sec)	Mean primary particle number per agglomerate	calculated primary particle emission rate	Median value of primary particle size distribution [nm]	GSD (average)
74 DX53	no wire	100% He	7,8	0,1	2,75	152	1,6	7,23E+11	477	3,5E+14	12	1,48
75 DX53	no wire	100% He	8	0,15	3,9	181	1,6	5,94E+11	2603	1,5E+15	13	1,5
76 AlSi1MgMn	no wire	100% He	2,6	0,25	5,8			2,14E+12	1375			
77 AlMg3Mn	no wire	100% He	6,5	0,15	2,6						18	1,5

Laser MIG hybrid welding

Pr. #	Parent metal	Wire	Gas	Mass emission rate [mg/s]	MMAD [µm]	GSD (average)	Number median of mobility distribution	GSD (average)	Agglomerate emission rate (1/sec)	Mean primary particle number per agglomerate	calculated primary particle emission rate	Median value of primary particle size distribution [nm]	GSD (average)
78 AlMg3Mn	AlMg5	50% Ar, 50% He	4,3	0,40	1,8	157	1,7	6,81E+11	366	2,5E+14	12	1,7
79 AlSi1MgMn	AlSi12	50% Ar, 50% He	1,2	0,40	2,7	97	1,7	9,96E+11	510	5,1E+14	13	1,6
80 DX53	Solid wire, low alloy	100% He	18,5	0,36	2,2	184	1,7	8E+11	202	1,6E+14	18	1,6
81 1.0038	Solid wire, low alloy	100% He	10,4			160	1,6	1,06E+12	2606	2,8E+15	20	1,4

Fig. 4.3 Characteristics of UFP in tungsten inert gas welding, laser welding and laser MIG hybrid welding. *Source* [6, 7]

Electron beam welding at atmosphere

Parent metal	Wire	Gas	Mass emission rate [mg/s]	MMAD [µm]	GSD (average)	Number median of mobility distribution	GSD (average)	Agglomerate emission rate (1/sec)	Mean primary particle number per agglomerate	Calculated primary particle emission rate	Median value of primary particle size distribution [nm]	GSD (average)
53 AlMg3Mn	no wire	He, air in addition	6,1	0,28	1,8	237	1,4	2,65E+12	138	3,7E+14	11	1,6
54 Mg–Al 3	no wire	He, air in addition	31,7	0,27	-	201	1,5	2,75E+12	704	1,9E+15	25	1,6
55 AlSi1MgMn	no wire	He, air in addition	22,2	0,17	1,9	124	1,5	3,65E+12	43	1,6E+14	19	1,7

Fig. 4.4 Characteristics of UFP in electron beam welding at atmosphere. *Source* [6, 7]

Resistance spot welding and resistance spot weld bonding

Pr. #	Parent metal	Wire	Gas	Mass emission rate [µg/s]	MMAD [µm]	GSD (average)	Number median of mobility distribution	GSD (average)	Agglomerate emission rate (1/sec)	Mean primary particle number per agglomerate	calculated primary particle emission rate	Median value of primary particle size distribution [nm]	GSD (average)
90	DX51 D+Z	no wire	no gas	23,2			47	2	8,20E+09	20	1,7E+11	14	1,6
91	DC01	no wire	no gas	37,7			54	2	4,55E+09	569	2,6E+12	12	1,4
92	organic coated	no wire	no gas	49,7	0,39	5,0			1,53E+11	201	3,1E+13	9	1,5
93	organic coated	no wire	no gas	50,1					2,11E+11	184	3,9E+13	10	1,5
94	DX51 D+Z	no wire	no gas	22,6						78		12	1,6
95	DC01	no wire	no gas	37,8					1,47E+10	123	1,8E+12	24	1,7

Fig. 4.5 Characteristics of UFP in resistance spot welding and spot weld bonding. *Source* [6, 7]

Soldering and brazing

Pr. #	Parent metal	Wire (Lot)	Gas	Mass emission rate [mg/s]	MMAD [µm]	GSD (average)	Number median of mobility distribution	GSD (average)	Agglomerate emission rate (1/sec)	Mean primary particle number per agglomerate	calculated primary particle emission rate	Median value of primary particle size distribution [nm]	GSD (average)
82.0090	Sn99Cu 1	no gas	1	0,45	2,6	58	1,5	3,25E+12				
83.0090	Sn95.5Ag3.8Cu0.7	no gas	0,8	0,56	4,3	65	1,5	4,35E+12				
84.1.8814	Ag55ZnCuSn	no gas	0,4	0,40	1,8	39	1,8	8,54E+11	12	1,0E+13	32	1,8
85.1.8814	Ag40CuZnSn	no gas	0,5	0,34	1,7	44	1,7	9,15E+11	9	8,4E+12	32	1,7

Fig. 4.6 Characteristics of UFP in soldering and brazing. *Source* [6, 7]

Plasma, flame, laser beam cutting

Pr. #	Parent metal	Wire	Gas	Mass emission rate [mg/s]	MMAD [μm]	GSD (average)	Number median of mobility distribution	GSD (average)	Agglomerate emission rate (1/sec)	Mean primary particle number per agglomerate	Calculated primary particle emission rate [1/sec]	Median value of primary particle size distribution [nm]	GSD (average)
86	1.4301	no wire	O2	0,66	0,12	2,1	172	1,6	1,1E+12	162	1,8E+14	9	1,5
87	1.0038	no wire	O2	0,21	0,35	11,9			4,75E+11	186	8,9E+13	8	1,5
88	1.4301	no wire	no gas	0,43	0,39	2,6			9,11E+10	202	1,8E+13	9	1,5
89	1.0038	no wire	no gas	0,21	0,12	3,0			2,58E+11	115	3,0E+13	7	1,4

Fig. 4.7 Characteristics of UFP in plasma, flame and laser beam cutting. *Source* [6, 7]

4.4 Findings from Research on Identification of Particle Characteristics in Welding of Galvanized Sheet

By means of the fume box method and on the basis of EN 15011, fire and electroplated hot dipped and electrolytic galvanized sheets were soldered with MIG-soldering and welded with, resistance spot welding and CO_2 laser beam welding in the ISF section of RWTH Aachen [8] in order to identify here again ultrafine particles and their particle characteristics.

The results of the measurement on emission rates related to the portion of particle fraction D < 100 nm, the particle number concentration and the particle surface concentration are shown in Fig. 4.8.

	MIG-soldering of galvanized sheets	Resistance spot welding of galvanized sheets	CO₂ laser welding of galvanized sheets
Average of mass emission rates	1,3 - 9,1 [mg/s]	0,07 - 0,18 [mg/s] 16,2 - 83,7 [µg/point]	2,56 - 4,97 [mg/s]
Average of portions of particle fraction D < 20 nm	8,2 - 12,8 %	39,05 - 78,65 %	9,00 - 12,11 %
D < 50 nm	10,65 - 16,78 %	81,56 - 95,88 %	12,99 - 14,98 %
D < 100 nm	26,49 - 55,14 %	96,66 - 99,55 %	40,38 - 55,16 %
Average of mean particle-number concentration	5,03E+06 - 2,56E+07 [particle/cm³]	5,70E+05 - 1,75E+06 [particle/cm³]	1,21E+07 - 2,22E+07 [particle/cm³]
Average of mean particle surface concentration	2,85E+12 - 2,61E+13 [nm²/cm³]	1,92E+10 - 1,12E+11 [nm²/cm³]	9,09E+12 - 1,50E+13 [nm²/cm³]
Average from agglomerate number emission rates identified from exhaust volume flow and particle number concentration	1,91E+11 - 9,73E+11 [particle/s]	1,51E+11 - 5,28E+11 [particle/point] 6,88E+11 - 1,89E+12 [particle/s welding current time]	4,59E+11 - 8,44E+11 [particle/s]

Fig. 4.8 Survey of particle characteristics measured. *Source* [8]

4.5 Other Findings

Experiments show, for processes with emission rates above 1 mg/s, a reduction of emission rates using targeted protective measures will provide a reduction in both the primary particle emission rate and the agglomerate emission rate [3].

In processes with medium, high and very high emission rates, the use of effective exhaust systems properly positioned to collect the welding fume at the point of generation, will greatly reduce the hazard of welding fume. This reduction may be sufficient to meet the specified limit values for a given workplace [3].

References

1. Spiegel-Ciobanu, V. E. *Ultrafine particles created by welding and similar procedures*. BIA-Report 07/2003e, HVGB, St. Augustin, pp. 151–162.
2. Spiegel-Ciobanu, V. E. (2013). *Schweißrauche Schweißen und Schneiden Wissen Kompakt* (Band 2). Düsseldorf: DVS Media.
3. Spiegel-Ciobanu, V. E. (2009). *Matrix zur Beurteilung der Schadstoffbelastung durch Schweißrauche, Band 3/2009: Aachener Berichte Fügetechnik* (Herausgeber Prof. Dr.-Ing. Reisgen). Shaker Verlag.
4. Spiegel-Ciobanu, V. E. (2012). Occupational health and safety regulations with regard to welding and assessment of the exposure to welding fumes and of their effect. *Welding and Cutting, 11*(1), 61.
5. Lehnert, M., Pesch, B., Lotz, A., et al. (2012). Exposure to inhalable, respirable, and ultrafine particles in welding fume. *Annals of Occupational Hygiene, 56*, 557–567.
6. Pohlmann, G., & Holzinger, K. (2008). *Vergleichende Untersuchungen bezüglich der Charakterisierung der ultrafeinen Partikeln in Schweißrauchen beim Schweißen und bei verwandten Verfahren*, Abschlussbericht 2008, Auftraggeber: Vereinigung der Metall-Berufsgenossenschaften (VMBG).
7. Pohlmann, G., Holzinger, K., & Spiegel-Ciobanu, V. E. (2013). Comparative investigations in order to characterise ultrafine particles in fumes in the case of welding and allied processes. *Welding and Cutting, 12*(2), 97–105.
8. Reisgen, U., Olschock, S., Lenz, K., & Spiegel-Ciobanu, V. E. (2012). Ermittlung von Schweißrauchdaten und Partikelkenngrößen bei verzinkten Werkstoffen [Determination of welding fume data and characteristic particle parameters in the case of galvanised materials]. *Schweißen und Schneiden, 64*, 788–798.

Chapter 5
Hazard Evaluation During Welding

To determine the health and safety program required to protect welder workers, an evaluation of workplace hazards which complies with national, regional or international health and safety regulations (e.g. labour protection law), shall be completed.

Knowledge of the "emission rate" and the "chemical composition" of the welding fume is a precondition for the hazard evaluation and for the choice of suitable protection measures for each process/material combination [1].

The information about the process and the material specific type of welding fume generated, the level of emission rate and the most important chemical components can be reached using the ISO 15011 "Laboratory method for sampling fumes and gases". Part 4 of this standard, "Fume data sheet" can be used as the first step in evaluating the welding fume limit value for a specific type of fume.

5.1 ISO 15011-4 Approach

ISO 15011-4 "Health and safety in welding and allied processes—Laboratory method for sampling fumes and gases—Part 4: Fume data Sheets" is the international standard specifying conditions under which fume is generated for the purpose of obtaining fume emission and chemical composition data for use in health and safety applications using standardized procedures. The fume data sheets are issued based on this data.

The standard also includes methods to evaluate exposure carrying out a gravimetric measurement of personal exposure to welding fume and is based on available welding fume data measured with laboratory methods. Results may be compared with a limit value that protects against the key component of a welding fume.

The welding fume limit value for the single element can be evaluated as follows:

$$LV_{WF(SC_i)} = \frac{LV_i}{i} \times 100$$

© International Institute of Welding 2020
V. E. Spiegel-Ciobanu et al., *Hazardous Substances in Welding and Allied Processes*, IIW Collection, https://doi.org/10.1007/978-3-030-36926-2_5

where

- $LV_{WF(SC_i)}$ is the single-component welding fume limit value calculated for the principal component of the fume, in mg/m^3, i.e. the welding fume concentration at which the limit value for the ith principal component of the fume is exceeded;
- LV_i is the limit value for the ith principal component of the welding fume;
- i is the proportion of the ith principal component of the welding fume, in % (mass fraction), as reported on the fume data sheet.

Results from gravimetric measurements of personal exposure shall be then compared with the lowest of these single-component welding fume limit values, i.e. the key-component welding fume limit value, $LV_{WF(KC)}$ to estimate whether welders are exposed to any component of the welding fume at a concentration in excess of its limit value. The method requires rounding the Key-component welding fume limit values the nearest 0.1 mg/m^3.

The above method may be used in case an additive welding fume limit value is not used to prescribe limits for complex substances that are mixture of chemical agents. The following approach is used in those countries that use risk assessment to ensure that the effects of the various components are at least additive, thus prescribing legal values:

$$LV_{WF(A)} = \frac{100}{\sum_1^n \frac{i}{LV_i} + \frac{(100 - \sum_1^n i)}{LV_{WF}}}$$

where

- $LV_{WF(A)}$ is the additive welding fume limit value in mg/m^3;
- LV_i is the limit value for the ith principal component of the welding fume;
- n the number of principal components of the welding fume;
- i is the proportion of the ith principal component of the welding fume, in % (mass fraction), as reported on the fume data sheet;
- LV_{WF} is the limit value, in mg/m^3, for welding fume containing only chemical agents of low to moderate toxicity, if such a limit has been set, or the limit value, in mg/m^3, for respirable fraction if no limit value for welding fume has been set.

The calculated additive welding fume limit value for the welding consumable in use shall be then compared with results from gravimetric measurements of personal exposure. The method requires to round additive welding fume limit value to the nearest 0.1 mg/m^3.

5.2   Risk Assessment

Based on information presented e.g. in fume data sheets for many types of electrodes/consumables during different welding processes, it is possible to evaluate the hazard and to make a risk assessment according to the level of the fume emission

(emission rate: e.g. low, middle, high, very high) and to the effects of the welding fume components (e.g. lung stressing, toxic, carcinognic).

The Appendix gives an example of risk assessment, based on the German approach [1, 2].

5.3 Assessment of Airborne Particles During Welding and Allied Processes

For all processes with *unalloyed/low alloy material (parent and filler materials), where the portions of chromium, nickel, cobalt, manganese, copper, barium, fluoride are individually below 5% by weight and* where no mutagenic, carcinogenic, fibrogenic, toxic or sensitising substances are contained in the welding fume, it is in most cases sufficient to determine the concentration of the welding fume/respirable fraction for comparison with the relevant limit value specified for workplace exposure.

The limit value for the respirable fraction can be the upper limit

During thermal spraying or for mixed workplaces (welding and grinding) the inhalable fraction can be important in addition to the respirable one. Here, the determination of the inhalable fraction concentration is also recommended.

During all processes with high-alloy materials (parent and filler material), where the individual portions of chromium, nickel, cobalt, manganese, copper, barium, fluoride are at least 5% by weight and where mutagenic, carcinogenic, fibrogenic, toxic or sensitising substances are contained in the welding fume, the limit values of the relevant key component shall be observed. For the specified key components see as well Figs. 3.2a–3.5, 3.8, 3.9 and 3.10 on pp. XX–XX of the present booklet.

The welding fume concentration is subject to compliance with the key component and therefore depends on the

(a) processes and materials,
(b) chemical composition of the welding fume,
(c) concentration of the key component in the welding fume and its limit value.

In this case, the allowable welding fume concentration will be very low and below the respirable dust limit value or welding fume limit value.

The limit value of the relevant key component will be the determining factor

For all processes, where compliance with the valid air limit value for welding fume or the respirable fraction of the dust is not possible—in spite of ventilation measures— in certain areas, such as confined spaces (e.g. boilers, containers, ship raised access floors) as well as other areas with low/insufficient air exchange, additional protection measures (e.g. organisational measures and use of personal protective equipment) are necessary.

5.4   Additional Tools for Hazard Evaluation

Other alternative or additional approaches to perform the hazard evaluation may take care of toxicity and carcinogenicity of hazardous substances and are reported as follows (e.g. Technical guideline on "Hazardous substances in welding and allied processes" (DGUV-I 209-017, English: version: (http://publikationen.dguv.de/dguv/udt_dguv_main.aspx?FDOCUID=23445. Software is also available to support the hazard evaluation such as (examples):

- 5x better [3],
- The software offered by the Berufsgenossenschaft Holz und Metall (BGHM—German Liability Social Accident Insurance Institution for the woodworking and metalworking sectors) on its website (http://www.bghm.de Webcode 802) [4],
- Potential determinants of exposure to respirable and inhalable welding fume: in: [5].

References

1. Spiegel-Ciobanu, V. E. (2009). *Matrix zur Beurteilung der Schadstoffbelastung durch Schweißrauche, Band 3/2009: Aachener Berichte Fügetechnik* (Herausgeber Prof. Dr.-Ing. Reisgen). Shaker Verlag.
2. Spiegel-Ciobanu, V. E. (2002). Bewertung der Gefährdung durch Schweißrauche und Schutzmaßnahmen. *Schweißen und Schneiden, 54*(Heft 2). Düsseldorf: DVS-Verlag.
3. 5x better: https://www.5xbeter.nl/site/nl.
4. BGHM-Software "Assessment on welding exposure": https://www.bghm.de, Webcode 802.
5. Weiss, T., Pesch, B., Lotz, A., et al. (2013). Levels and predictors of airborne and internal exposure to chromium and nickel among welders—Results of the WELDOX study. *International Journal of Hygiene and Environmental Health, 216*, 175–183.

Chapter 6
Measurement Methods

Measurement of hazardous substances at the workplace provides a picture of the actual situation regarding hazardous substances, to allow appropriate decisions to be made on protective measures. These are aimed at assessing the concentration of hazardous substances present in the welder's breathing zone.

Different measurement strategies have been developed over the course of time, differing in cost and effort according to the level of confidence required for the result.

For this reason, testing measurements should be carefully planned, prepared and carried out in such a way that the result really represents the situation. For the same reason, it is of fundamental importance that a preliminary analysis of the processes and production must be completed.

Measurement methods for hazardous substances at the workplace usually consist of the following individual steps:

- arrangement for sampling
- sampling
- handling and storing of the sample
- preparation of the sample
- analytical determination
- calculation of the result.

The level of emissions of hazardous substances (mg/m^3) at the workplace and the corresponding exposure experienced by the welder are determined by different measurement methods. These are strongly affected by the sample carrier chosen based on the nature of the emission to be determined. Consistently, measurement procedures can be differentiated into:

- Measurement methods for gaseous substances (gases such as ozone, nitrogen oxides, carbon monoxide, etc.);
- Measurement methods for airborne particles (fume and dusts).

Different sampling procedures may be used; these are standardized by different bodies. The following paragraphs refer specifically to ISO 10882 "Health and safety

© International Institute of Welding 2020
V. E. Spiegel-Ciobanu et al., *Hazardous Substances in Welding and Allied Processes*, IIW Collection, https://doi.org/10.1007/978-3-030-36926-2_6

in welding and allied processes—Sampling of airborne particles and gases in the operator's breathing zone".

6.1 Measurement Methods for Gaseous Substances

Personal exposure may be determined through continuous measurement methods with direct reading measurement instruments (direct reading electrical apparatus and detector tubes, Fig. 6.1) or discontinuous measurement methods. Figure 6.1 shows different measurement methods according to international standard.

The most common method makes use of direct reading electrical apparatus as they can be more easily calibrated and can provide instantaneous measurements to indicate the amount and time of potential exposures to hazardous substances right at the workplace. This equipment are useful for screening measurements of either the variation of concentration in time, time weighted average concentration and for comparison with limit values and periodic measurements. This equipment requires qualified personnel for both the use and maintenance. Attention shall be paid to cross sensitivities.

Another possibility is using detector tubes, where a defined volume of air is aspirated from the open detector tube by a suitable manual or battery-driven pump (Fig. 6.2). The concentration can be read from the change in colour of the filler compound, which is specific for a hazardous substance or a group of substances. There are short and long term detector tubes. Detector tubes are most useful for screening measurements of time weighted average concentration rather than for comparison with limit values or periodic measurements.

Other methods require laboratory analysis, as sampling is made at the workplace and the evaluation is made at the laboratory. These include a system where the sampled air is drawn by a motor-driven pump through a suitable sorption tube (e.g. activated carbon or silica gel tubes) with adequate absorbing capacity and methods with a diffusive sampler without suction unit, where the sampling effectiveness

Measurement method	Gases and vapours				
	ozone (O_3)	carbon monoxide CO	carbon dioxide CO_2	nitrogen oxide (NO)+ nitrogen dioxide (NO_2)	organic vapours
	0,01 ppm to 3 ppm	3 ppm to 500 ppm	500 ppm to 10 %	0,3 ppm up to 250 ppm	-
Direct reading electrical apparatus	generally used	generally used	generally used	generally used	available, but usefulness limited by poor specificity
Detector tubes	available, but not recommended	generally used	generally used	generally used	available, but usefulness limited by poor specificity
indirect methods involving laboratory analysis	not generally applicable	not generally applicable	not generally applicable	available but not generally used	generally used
ISO 10882-2 Sampling of airborne particles and gases in the operator's breathing zone - Part 2: Sampling of gases					

Fig. 6.1 Measurement of individual gases and organic vapours. Courtesy ISO (Excerpt from ISO 10882-2; Annex B)

Fig. 6.2 Direct reading detector tubes: accuro 2000 one hand-operated gas detector pump and automatic pump with detector tube for the measurement of gaseous hazardous substances. *Source* Courtesy Draeger, Lübeck, Germany

may depend on the air motion (the welder breathing is generally to be considered sufficient).

6.2 Measurement Methods for Particulate Substances in the Breathing Zone

Sampling on the person with the personal dust collector (Personal Air Sampler = PAS) is carried out in the welder's breathing zone (Figs. 6.3 and 6.4) with a "sampling head" for the inhalable fraction, which collects airborne particles on a filter.

With the aid of a pump, attached to the welder's body by a belt, a certain volume of air is aspirated. According to ISO 10882-1 the sampler shall be located behind the welder's face shield (if applicable). It can have different positions: to the left or right of the face or under the chin. The measurement is carried out by the sampling head for the Inhalable fraction of the dust (Figs. 6.5, 6.6 and 6.7).

After sampling and transport of the filter to the laboratory, a quantitative (mg/m^3) and qualitative (chemical composition) analysis of the sample is made in the laboratory. The analytical determination by weighing and chemical analysis usually is restricted to the key components.

In addition to the inhalable fraction, devices are available to measure also the respirable fraction. This is the case of the "PGP-EA" used for the determination of the mass concentrations of respirable and inhalable fraction and the substance

Fig. 6.3 Measurement of
airborne particles with a
personal dust collector (PAS)
and check for gaseous
hazardous substances by
means of detector tubes

Fig. 6.4 Personal
measurement with PAS for
the determination of
particulate hazardous
substances. *Source*
Spiegel-Ciobanu [1]

Fig. 6.5 Welder's face
shield with a sampler
attached by means of a
removable clip. *Source*
Courtesy ISO 10882-1

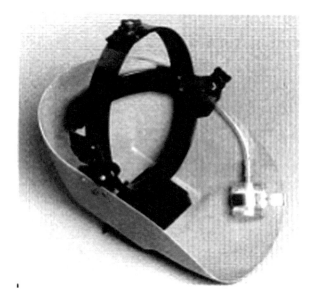

Fig. 6.6 Welder wearing a
sampler attached to a
sportsperson's headband.
Source Courtesy ISO
10882-1

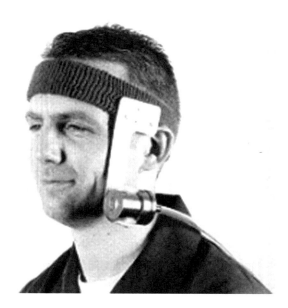

concentrations in the respirable dust during welding and allied processes as a standard system in the measuring system of the Berufsgenossenschaften for hazardous substances (BGMG).

In analogy to the sampling head GSP 3.5 the inhalable fraction is captured by a suction cone with a volume flow of 3.5 l/min. Inside the sampling head PGP-EA the

Fig. 6.7 Welder wearing a sampler attached to a sportsperson's headband and a welder's face shield. *Source* Courtesy ISO 10882-1

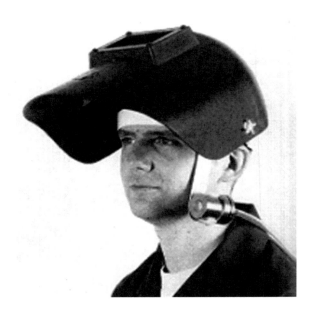

aerosol is divided into two fractions. For this purpose, an open pore polyurethane foam as size selective and collecting element is used together with a downstream plane filter [2, 3]. The porosity and the geometry of the polyurethane foam are chosen such that respirable fraction particles according to the definition in EN 481 may pass. The foam is inserted into an annular holder, the plane filter with a diameter of 37 mm is inserted into the usual capsule in connection to the universal mounting device of the PGP system (see Figs. 6.8 and 6.9). Both filter elements shall be transported together in a tin, inserted into the sampling head on site and mounted with the special suction cone to the filter mounting device of the PGP system.

The inhalable fraction concentration is determined by weighing both filter elements (foam and plane filter) and the respirable fraction by weighing only the plane filter. The elementary analysis of dust in the plane filter is possible by means on standard methods [2] (Fig. 6.10).

6.3   Stationary Sampling for Particulate Substances

Sampling with a stationary dust collector (Fig. 6.11) is made at a fixed point in the vicinity of the welder (working area). The sampling position in the room is chosen such that it is suitable for determining the general concentration of welding fume in the workplace atmosphere. It is also used for evaluating the ventilation conditions in the room.

Procedures for the assessment and evaluation of collected data are depending on the composition of the particles.

Fig. 6.8 Sampling head PGP-EA. *Source* DGUV I 209-017, Courtesy Institute for Occupational Safety and Health of the German Social Accident Insurance (IFA)

Fig. 6.9 Dismounted sampling head PGP-EA. *Source* DGUV I 209-017, Courtesy Institute for Occupational Safety and Health of the German Social Accident Insurance (IFA)

In all cases where there is not any significant content in key components it may be sufficient determining the concentration of the welding fume/respirable fraction to compare it with the relevant limit value. Generally, respirable fraction/welding fume is the upper limit and is decisive. However, during thermal spraying, thermal cutting or for mixed workplaces (welding and grinding) the inhalable fraction can also be significant.

In all other cases (e.g. welding *high-alloy materials*) and when key components are contained in the welding fume, the limit values of the relevant key component shall be observed, even if the limit value for the respirable fraction is not exceeded. Here, the chemical composition of the welding fume, the concentration of the key component in the welding fume and its limit value are very important.

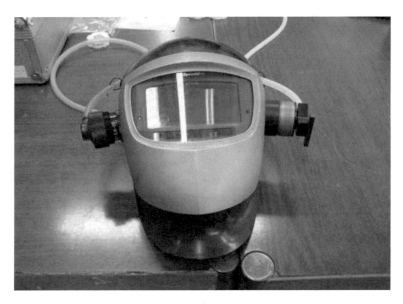

Fig. 6.10 Welder's safety helmet with integrated sampling heads PGP-EA and GSP for simultaneous sampling. *Source* Spiegel-Ciobanu [1]

Fig. 6.11 Stationary dust or
fume collector (VC 25) for
fixed point sampling
(measurement) of particulate
hazardous substances.
Source Spiegel-Ciobanu [1]

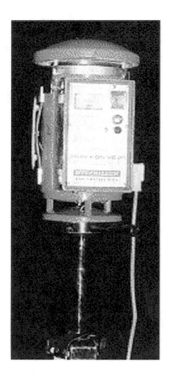

As a result of the analysis, when compliance with the valid air limit value for welding fume or the respirable fraction of the inhalable fraction is not possible—in spite of ventilation measures- in certain areas, such as confined spaces (e.g. boilers, containers, ship raised access floors) as well as other areas with low/insufficient air exchange, additional protection measures (e.g. organisational measures and use of personal protective equipment) are necessary.

References

1. Spiegel-Ciobanu, V. E. *Hazardous substances in welding and allied processes*. German Social Accident Insurance (DGUV), Information 209-017.
2. Möhlmann, C., Aitken, R. J., Kenny, L. C., Görner, P., VuDuc, T., & Zimbelli, G. (2003, October). Größenselektive personenbezogene Staubprobenahme: Verwendung offenporiger Schäume. *Gefahrstoffe – Reinhaltung der Luft, 63*(10), 413–416.
3. Kenny, L. C., Aitken, J. J., & Görner, P. (2001). Investigation and application of a model for porous foam aerosol penetration. *Journal of Aerosol Science, 32,* 271–285.

Chapter 7
Technical Protective Measures

In order to minimise the health hazard to the welder by hazardous substances at the workplace, technical, organisational and—in certain situations—also personal protective measures shall be implemented according to the applicable regulations. The efficiency of the implemented protective measures shall be checked by verification of their efficiency.

The technical protective measures listed in the following shall be chosen individually or in combination.

7.1 Selection of Low Fume Emission Processes

Welding is a very complex technology, and the same product may be obtained by using different processes. When designing the fabrication process, fume emission shall be taken into consideration together with costs, productivity, etc.

As an example, when considering arc welding processes, the following aspects may be taken into account:

- In general, welding parameters shall be optimized to reduce fume emission.
- Manual Metal Arc electrodes are often used in welding of chromium-nickel steel. This process may produce fume which contain hexavalent chromium compounds in critical concentrations. In some situation it would be possible considering converting to metal active gas welding (MAG). Although MAG welding produces more chromium compounds overall, these are, however, predominantly trivalent [1, 2].
- In comparison to manual metal arc welding, metal active gas welding (MAG) and metal inert gas welding (MIG), tungsten inert gas welding (TIG) generates much less fume (Fig. 7.1).
- Shielding gas composition and transfer modes may significantly affect the Fume Emission Rate in GMAW/FCAW.

© International Institute of Welding 2020
V. E. Spiegel-Ciobanu et al., *Hazardous Substances in Welding and Allied Processes*, IIW Collection, https://doi.org/10.1007/978-3-030-36926-2_7

Fig. 7.1 TIG welding, a process with low welding fume emission. *Source* Spiegel-Ciobanu [3]

- In submerged arc welding the welding process is carried out under a blanket of granular flux (Fig. 7.2). Thus, only small quantities of hazardous substances are emitted. In addition, for procedural reasons, the operator is generally not very close to the weld. For these reasons, submerged arc welding is recommended as an alternative to other arc welding processes—if technically possible.
- In MIG/MAG, wave-form controlled processes (e.g. ColdArc®, CMT®, STT®, RMD®, etc.) may lead to a reduction of the welding fume emission up to 75–90% [4].

Fig. 7.2 Submerged arc welding with low emission of hazardous substances. *Source* Spiegel-Ciobanu [3]

– Laser flame cutting of unalloyed and low-alloy steel leads to lower emissions of hazardous substances than flame cutting.
– Laser beam high pressure cutting (with N_2) instead of laser flame cutting (with O_2). For the same material and the same sheet thickness, emissions of hazardous substances during high pressure laser cutting can be lower by a factor of 2–15 than those during laser flame cutting.
– Where possible, flame spraying should be preferred to arc spraying, because of the lower fume emissions. On the other hand, the risk of generation of nitrous gases shall not be neglected here.

7.2   Selection of Low Emission Consumables

When considering arc welding processes, the following aspects related to the welding consumables shall be considered:

– Individual consumables in available welding processes will have different fume emission rates. Additionally, different grades may generate different emission rates. Some coating types, for instance SMAW stainless steel electrodes, may have reduced hexavalent chromium due to different silicate binders or due to applied core wires.
– In order to lower the fume emission rate as well as the hazardous compounds in it, different processes may be applied, provided productivity, efficiency and technical soundness are met.
– In general, the fume emission rate (FER) increases from SAW, TIG, MIG, MAG, FCAW/SMAW.
– Choice of shielding gas may affect the FER.

7.3   Improvement of the Working Conditions

The generation of welding fume, including hazardous substances, can be reduced by selection of favourable welding parameters. Additionally, improvement of working conditions as well as ventilation and PPE will contribute to a lower exposure.

• The selection of favourable welding parameters may contribute substantially to minimising hazardous substances. It is advisable to avoid high levels of welding voltage, welding current and shielding gas flow rate. Please contact supplier for optimal parameters.
• Non-thoriated electrodes for TIG welding with other oxide additions (e.g. additions containing cerium or lanthanum) are already standardized and available on the market.

- In laser cladding, hazardous substances can be minimised by lowest possible energy per added powder in relation to the process result and best possible powder selection with respect to particle size distribution.
- In cutting, fumes emission and hazardous substances can be minimised. Contact suppliers for optimal parameters
- During welding and cutting operations on coated workpieces, additional hazardous substances are generated from the coating which can be avoided by the following measures:

  – reduction of the coating thickness to as low as practically feasible,
  – removal of coatings in the welding area,
  – removal of contamination on the workpiece surface (e.g. oil, grease, paint, residues of solvents).

- The welder's workplace and the positioning of the workpiece should be such that hazardous substances ascending with the thermal updraft can largely be kept away from the welder's breathing zone, as follows:

  – the horizontal distance between the welding location and the welder's head is as large as possible and
  – the vertical distance is as short as possible.

7.4   Technical Safety Devices

Some processes may allow special technical safety devices to reduce the emission of hazardous substances in the environment.

Some examples are reported as follows.

7.4.1   Oxyacetylene Welding and Cutting

At fixed workplaces for oxyacetylene operations it is possible to use torch holders fitted with automatic gas shut-off valves. Similarly, some torches have a system to shut-off gas and oxygen outlet when not in use, with a simple ignitor to re-start operations.

This avoids the generation of large quantities of nitrous gases during breaks.

Studies on flame cutting with a water shielding or under water show a significant reduction in the emission of hazardous substances compared to atmospheric flame cutting:

– particulate emissions are reduced by several orders of magnitude,
– emission of gaseous substances also decreases substantially.

7.4.2 Plasma Cutting

Plasma cutting with a water curtain/air-water shower or water protection bell water curtain is usually carried out in conjunction with a water cutting table and a water injection cutting torch. The emission of hazardous substances is reduced but cannot be avoided.

Water shielded plasma cutting reduces the emission of hazardous substances and noise to a considerable extent. Depending on the application (sheet thickness, type of material), emission of airborne particles may be reduced by a factor of up to 500 for comparable cutting operations. The emission of gases, especially nitrous gases (in plasma cutting with argon/nitrogen/hydrogen as plasma gas), may be reduced up to a half.

By underwater plasma cutting (i.e. cutting on the water surface) the sheet is placed on the water surface of a cutting tank and a concentric exhaust can be installed around the torch so that the emission of hazardous substances is reduced.

A comparison between plasma cutting in atmosphere, plasma cutting above water bath and plasma cutting under water shows the following reduction in emission:

$$\text{for dust} = 100{:}10{:}1$$
$$\text{for NO}_x = 4{:}2{:}1$$

7.4.3 Operations in Enclosed Booths

Operating welding, cutting and allied processes in enclosed booth may eliminate any exposure when a welder is controlling the process externally (automatic processes).

This is nowadays very common in thermal spraying (particularly with plasma spraying) and frequent with laser processing (Fig. 7.3).

7.4.4 Ventilation

Practical experience shows that ventilation often is the only possibility for minimising hazardous substances.

7.4.5 Free Ventilation (Natural Ventilation)

Air is exchanged between the interior and the exterior as a result of a pressure gradient due to wind and temperature differences between the outside and the inside.

Fig. 7.3 Enclosed booth for plasma spraying. *Source* Spiegel-Ciobanu [3]

The exchange of air takes place through windows, doors, ridge turrets etc. Free/General ventilation can only be the solution for low hazard levels (determined by amount, concentration, type of hazardous substance).

An example is TIG welding with non-thoriated tungsten electrodes.

7.4.6 Forced (Mechanical) Ventilation

The exchange of air between the room and the outside realized by circulating units (e.g. fans or blowers) is called forced (mechanical) ventilation.

To obtain effective forced ventilation in rooms or halls, the air flow pattern, for example in welding—thermal uplift of hazardous substances from the bottom to the top—should be arranged so that the exhaust air is removed from the upper part of the room and fresh air flows in at the bottom.

7.4.7 Extraction

In welding processes where high concentrations of hazardous substances and/or critical substances are to be expected in the workplace atmosphere, the use of extraction systems (Fig. 7.4) for a direct local capture of the generated hazardous substances is the most effective protective measure.

All process/material combinations known to cause medium, high and very high hazards are addressed.

Fig. 7.4 Example of a stationary extraction. Courtesy of Fa. KEMPER, Vreden, Germany

Stationary extraction equipment is suitable for repetitive welding operations carried out at fixed locations (e.g. mass production), while mobile fume extraction devices and or portable devices are commonly more used in those situations where welding operations are performed in different work-areas.

The purpose is optimum capture and exhaust of hazardous substances and high efficiency of the filter system.

The capture element is determining for the effectiveness of extraction (Fig. 7.5).

The choice of shape, the correct dimensioning and positioning of capture elements shall correspond to the thermal movement of the welding fume and their quantity and depends on the working situation.

The capture element should always be placed as close as possible to the generation area of hazardous substances.

For flexible capture elements, it is important that the welder is willing to position it correctly.

Capture elements with a flange are shown to be more effective than the conventional types without flanges used previously. Tests on a test stand show that refitting funnel with a flange of a width of 50 mm may lead to an increase of suction reach of about 10% in all directions or to a reduction of air quantity demand of about 20%.

The filter system used plays a decisive role in the separation of hazardous substances. Among other factors, the choice of these filter systems depends on the

Fig. 7.5 Gas-shielded metal arc welding (MIG/MAG) with extractor. Courtesy of Fa. KEMPER, Vreden, Germany

chemical composition of the hazardous substances. For the filtration of metal dusts, self-cleaning surface filters can be recommended.

Separation of gases and especially of organic components is very difficult and shall be adapted to each single case (process, material). The extraction equipment requires an efficient separation of the hazardous substances for the recirculation of the air and/or for the environment.

Various filter systems—mechanical, electrostatic—are available.

Available **extraction equipment** may be differentiated into:

- static extraction equipment and
- mobile extraction and or portable devices.

Static extraction equipment is suitable for repetitive welding operations carried out at fixed locations (e.g. mass production). Extracted air is directed to the outside by ducting (e.g. within a larger centralised system).

Depending on the task, the capture hood is fixed in position or can be guided by flexible hoses.

These units are available in different forms.

7.4.8   Downdraft Cutting Tables

Cutting tables are usually fitted with down-draught extraction (extraction below the table). Down-draught extraction acts in the direction opposite to the thermal up-current of hazardous substances. An additional extraction device can be fitted above the table and to the rear. Extractor tables fitted to flame and plasma cutting systems are intended to work in sections and hence concentrate on the respective dust generation area.

Booths with extraction for welding and thermal spraying

Booths incorporating extraction have yielded very good results in practice. The walls of the booth enclose the working area as closely as possible. Air flowing in through the booth opening is exhausted on the opposite side. Inside these booths, hazardous substances should be captured and extracted directly at the area of generation by special capture devices. It may be necessary when entering the booth area, to wear Persoanal Respiratory Protection (PRP), or Personal Protective Equipment, (PEP), when fumes and particulate cannot be adequately controlled using ventilation techniques.

In general, experimental results show that extraction pipes with flanges provide improved capture performance.

In laser processes booths also offer a good protection against hazardous substances and optic radiation.

The separator may be connected to the central extraction system or the booth may have its own separator.

Mobile welding fume extraction devices

The devices may be suitable when the workplaces vary and can be used in many fields. These devices operate with air recirculation, i.e. the captured and filtered air is recirculated into the working area. Attention should be paid to gaseous substances (appropriate filtering shall be used).

Use of these devices may be restricted based on applicable law or regulations.

Mobile extraction devices may be combined and operated with different capture devices:

- gas shielded torches with integrated extraction,
- hand shield with integrated extraction
- movable capture devices (extractor ducts).

Fig. 7.6 Capture element
fixed to the torch. Courtesy
of DINSE GmbH, Hamburg,
Germany

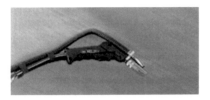

Fig. 7.7 Torch with
integrated extraction.
Courtesy of DINSE GmbH,
Hamburg, Germany

Torch welding fume extracting device

Formerly, the extraction and its regulation were done by means of an annular nozzle; the new version, where the extraction device is attached to the torch so that it can be turned, is better adaptable to the local conditions and the welding situation. A larger degree of capture is reached without noteworthy adverse effect on the gas shielding.

Existing torches can also be refitted with it.

For recirculation systems, the proportion of the recirculated air shall not exceed 50%.

Ventilation systems with recirculation may be used when they are type tested or when their required efficiency has been verified.

The proportion of non-extracted hazardous substances is decisive in case less than 85% are extracted.

Today, mobile filter extractors to the state of the art are successfully used for welding (Fig. 7.6).

Torches with integrated extraction are suitable for MIG/MAG processes. They allow direct extraction at the point of generation of hazardous substances. They may be used in stationary and in mobile systems (Fig. 7.7).

For Nd:YAG lasers working head integrated extraction nozzles were also developed (Fig. 7.8).

When designed for high vacuum, capture elements and piping/hoses only need small diameters as the air volumes required for ventilation are small.

7.4.9 Supply Air Systems/Room Ventilation

Generally, for welding processes in halls supply air and exhaust air systems are required for the removal of hazardous substances. The air contaminated with hazardous substances shall be suitably replaced by uncontaminated air. The general air

Fig. 7.8 Working head
integrated extraction for
Nd:YAG lasers. Courtesy of
Heinz-Piest-Institut,
Hannover, Germany, PHI
extractor nozzle MK 5

cleaning method must create enough air changes per hour to overcome the amount
of contaminant being generated. Depending on the type of welding/operation being
performed, 10 to 25 or more air changes per hour may be required to maintain desired
levels of contaminant removal.

For process/material combinations, where a low hazard can be anticipated as e.g.
submerged arc welding a technical room ventilation can generally be sufficient for
welding activities taking a longer time.

References

1. Spiegel-Ciobanu, V. E. (1999). Beurteilung partikelförmiger Stoffe in der Schweißtechnik.
   Schweißen und Schneiden, 51(Heft 4). Düsseldorf: DVS-Verlag.
2. Spiegel-Ciobanu, V. E. *Welding activities with chromium and nickel alloyed filler and base metal,
   DGUV-Information 209-059 BGI 855E.*
3. Spiegel-Ciobanu, V. E. *Hazardous substances in welding and allied processes.* German Social
   Accident Insurance (DGUV), Information 209-017.
4. Rose, S. Ansätze zur Entstehung und Reduzierung von Schweißrauchemissionen beim MSG-
   Schweißen unter Berücksichtigung neuer Verfahrensvarianten - Ergebnisse des 1. EWM17
   Awards, Physics of Welding. *Schweißen und Schneiden* 04/2012.

Chapter 8
Personal Protective Equipment (PPE) for Hazardous Substances

To minimise the health hazard to the welder posed by hazardous substances encountered in the workplace, engineering, administrative organisational and—in certain situations—personal protective measures, shall be implemented. The effectiveness of the implemented protective measures shall be verified.

Personal protective equipment is intended to protect the welder directly and, in many cases, is a necessary supplement to the engineering and administrative measures.

8.1 Welder's Hand-Held Shields and Welding Helmets

Hand-held shields or welding helmets with appropriate filter lenses shall be used in arc welding. They provide protection from optic radiation, heat, sparks and, to some degree, from hazardous substances. The welder is responsible for their correct positioning.

8.2 Respiratory Protective Equipment

When the ventilation measures used, in the area where hazardous substances are generated, do not reduce exposure levels below limit values, or when concentrations of carcinogenic substances are not sufficiently minimized, respiratory protective equipment shall be used as an additional measure. See here Fig. 8.1.

Furthermore, the use of respiratory protective devices is only allowed if all administrative and engineering measures have been found to be insufficient to protect the welder. In general, respiratory protection devices should only be used for short periods or in confined spaces (e.g. boilers, containers, raised access floor cells in ships) or other areas with low/insufficient air exchange.

© International Institute of Welding 2020
V. E. Spiegel-Ciobanu et al., *Hazardous Substances in Welding and Allied Processes*, IIW Collection, https://doi.org/10.1007/978-3-030-36926-2_8

Fig. 8.1 Welding with respiratory protection during works in constraint posture. Courtesy 3M Deutschland GmbH, Kleinostheim, Germany

For certain process/material combinations (e.g. MIG welding of aluminium materials), we know from experience that minimizing the hazardous substances concentration—by means of ventilation measures—does not suffice to achieve that both the concentrations of ozone and welding fume remain under the limit values. In these cases, a Powered Air-Purifying Respirator (PAPR) helmet, is recommended as a supplement to ventilation.

Even for low emission processes where welding fume contains carcinogenic substances such as Cr(VI) or NiO the residual risk in the welder's breathing zone shall be reduced by wearing respiratory protective equipment.

Chapter 9
Preventive Occupational Medical Care

Despite considerable technical and organisational efforts, full protection from the effects of hazardous substances cannot be attained under today's operating conditions. Therefore, it is necessary and beyond dispute that, for some operations, necessary technical prevention should be accompanied by preventive occupational medical care. The purpose is avoidance and early detection especially of chronic diseases.

Employers shall follow the respective national regulations on preventive occupational medical care and surveillance for welders.

Regulations vary from country to country. The remarks made here only provide a guideline for the interview and examination of welders by medical staff.

Main potential target organs of welding fumes and gases

The sites and organs mainly affected by fumes and gases from welding, and allied processes are the **respiratory tract and the lungs**.

The following **effects/diseases** may occur if fume and/or gas exposures exceed safe values. Pre-existing non-occupational diseases can be enhanced by exposure to welding fumes and gases [1–4]:

- Irritation of the upper airways (e.g. due to ozone)
- Dyspnoea and fever (metal fume fever e.g. due ZnO and CuO)
- Chronic bronchitis
- Obstructive airways disease, asthma (e.g. due to ozone, thermal decomposition products of organic layers, fluxes)
- Chronic obstructive lung/pulmonary disease (COPD)
- Bacterial infections of the lungs, mainly by pneumococcus (pneumococcal pneumonia)
- Inflammation of the lung (alveolitis, pneumonitis), lung fibrosis (siderofibrosis, aluminosis: rare cases)
- Lung cancer (there is a slight but not neglectable excess risk for arc welders).

© International Institute of Welding 2020
V. E. Spiegel-Ciobanu et al., *Hazardous Substances in Welding and Allied Processes*, IIW Collection, https://doi.org/10.1007/978-3-030-36926-2_9

The **following symptoms should be checked** when taking the medical history of welders. It is noteworthy that the latency time between the start of the fume/gas exposure in the work shift and the development of acute or chronic symptoms may vary, depending on the amount and the content/components of the fumes/gases on the one hand and on the underlying disease on the other hand:

– Metallic taste (in case of metal fume fever e.g. due ZnO and CuO)
– Fever, chill
– Burning or running nose
– Sneezing
– Cough
– Expectoration of phlegm/sputum
– Wheezing, rhonchi
– Shortness of breath, dyspnoea (at rest and/or at exercise)
– Loss of weight (can be a non-specific symptom of lung cancer).

The **Medical examination** should include at least:

• Physical examination: inspection of the conjunctivae and the throat; auscultation of the lungs
• Lung function (1)

   – as a minimum: spirometry including at least FEV and FEV_1.

• if possible flow-volume-curves.

   – comparison of lung function with previous results,
   – if warranted: diffusion measurements, blood gas analyses, (spiro)ergometry and,

• Biological Monitoring where appropriate (2) [4].
  Looking to identify, Chromium, Nickel, Aluminium in urine; and other metals identified in the welding plume,
• Chest X-ray, computed tomography of the lungs when the symptoms are unclear.

It shall be noted that tobacco smoking is a strong confounding factor for airways and lung diseases.

(1) Lung function tests must be performed per the national and international guidelines of the pneumological medical associations. This is important for quality reasons and for the validity of the comparison with previous lung function tests of the welder. A loss of more than 50 ml FEV_1 per year should be cause for concern as it is about double the loss that can be regularly expected from aging.
(2) Biological monitoring is a valuable tool in occupational medicine (in addition to air monitoring) for the assessment of the intake of metals such as chromium, chromium VI, nickel, aluminium and other metals, when these are constituents of the fume. In contrast to air monitoring, it better reflects the individual situation in the breathing zone and it also regards the individual breathing volume (triggered by the degree of exercise at work). Only accredited laboratories which have quality control and experience in this very specialised field should be used

for the analysis. Further information is given for instance in the recommendations of Biological Exposure Indices (BEI, USA) and the Biological Exposure Tolerance Values und Exposure equivalent values for carcinogenic substances (BAT, EKA, Germany) [5, 6].

In individual cases, neurological examinations may be indicated, if there is a high exposure to potentially neurotoxic fume components such as Manganese. It should be noticed that—besides manganism, which is currently found in welders only in exceptional cases—the question, whether workers can develop neurological symptoms from welding is unclear so far.

References

1. Cosgrove, M. (2015). Arc welding and airway disease. *Welding in the World, 59,* 1–7.
2. Cosgrove, M. P., & Zschiesche, W. (2016). Arc welding of steels and pulmonary fibrosis. *Welding in the World, 60,* 191–199.
3. Palmer, K. T., & Cosgrove, M. P. (2012). Vaccinating welders against pneumonia. *Occupational Medicine, 62,* 325–330.
4. Weiss, T., Pesch, B., Lotz, A., et al. (2013). Levels and predictors of airborne and internal exposure to chromium and nickel among welders—results of the WELDOX study. *International Journal of Hygiene and Environmental Health, 216,* 175–183.
5. German Research Foundation (DFG). *List of MAK and BAT values 2018.* Wiley Online Library: https://onlinelibrary.wiley.com/doi/book/10.1002/9783527818402 and http://www.dfg.de/dfg_profil/gremien/senat/arbeitsstoffe/.
6. American Conference of Governmental Industrial Hygienists (ACGIH): Biological Exposure Indices (BEI®). http://www.acgih.org/tlv-bei-guidelines/biological-exposure-indices-introduction.

Appendix
Hazard evaluation during welding, German approach

As a global approach, ISO 15011-4 "Health and safety in welding and allied processes—laboratory method for sampling fumes and gases. Part 4: Fume data sheets" includes an Example of classification of welding consumables in its appendix F; further information can be found in Chap. 5 of this book.

This appendix includes, as a further example, an excerpt from the DGUV-Information booklet 209-016 "Hazardous substances in welding and allied processes", Edition 2012, Author: Vilia Elena Spiegel-Ciobanu, edited by BGHM, Mainz, Germany, based on [1].

The following factors are part of the hazard evaluation:

Process specific factors

- The welding processes may be classified into four classes according to emission rates (mg/s) (emission classes 1–4) with respect to particles.

Effect specific factors

- The welding fume may be classified into three classes (effect classes A, B, C) with respect to the specific effect of their components on the human body.

The extent of health hazard (low up to very high) depends on process-specific and effect-specific factors (Fig. A.1).

Figure A.2 on page 90 contains an assignment to the welding fume classes based on emission rates and effect (A1 to C4).

Workplace specific factors

- They include especially: spatial conditions, ventilation situation, head and body position during welding.

Hazard	Welding fume category
I low health hazard	A1
II medium health hazard	A2, B1, C1
III high health hazard	A3, B2, B3, C2, C3
IV very high health hazard	A4, B4, C4

Fig. A.1 Classification of the hazards into the welding fume categories

© International Institute of Welding 2020
V. E. Spiegel-Ciobanu et al., *Hazardous Substances in Welding and Allied Processes*, IIW Collection, https://doi.org/10.1007/978-3-030-36926-2

Welding fume: Emission classes/ emission rates [mg/s]	Process examples	Welding fume: effect		
		Effect class A	Effect class B	Effect class C
		Substances straining respiratory tract and lung[1])	Toxic or toxic irritating substances[2])	Carcinogenic substances[2])
		e.g. Fe_2O_3	e.g. F^-, MnO, CuO	e.g. Cr(VI), NiO
		Hazard	Hazard	Hazard
1	< 1	e.g. UP[3])	I (A1)	I (B1)	I (C1)
	< 1	e.g. TIG[4])	I (A1)	II (B1)	II (C1)
2	1 to 2	e.g. laser welding	II (A2)	III (B2)	III (C2)
3	2 to 25	e.g. MMA, MAG (solid wire)	III (A3)	III (B3)	III (C3)
4	> 25	e.g. MAG	IV (A4)	IV (B4)	IV (C4)

I = low health hazard   II = medium health hazard   III = high health hazard
IV = very high health hazard;

A1 to C4: welding fume classes

[1]   If alloying or cover/filler components each are < 5%.

[2]   If alloying or cover/filler components each are > 5%.

[3]   Automated

[4]   See Ref. [2,3]

Fig. A.2 Hazard evaluation on the basis of emission rates and effect; classification into the welding fume classes

For **medium, high and very high emission rates (emission classes 2 to 4)**—without ventilation measures-, **concentrations of hazardous substances** occur in the breathing zone of the welder, which **exceed the limit values by far**.

For **low emission rates (emission class 1)**, the concentrations of hazardous substances in the breathing zone of the welder are by experience known to be on the level of the limit value or slightly below.

Without ventilation measures and due to additional circumstances at the workplace the health hazard is in some cases increased, e.g. in confined spaces.

Therefore, besides considerations concerning

1. selection of low emission processes
2. selection of low emission materials

best possible solutions for ventilation—as far as technically possible—shall as well be found and used.

It may be assumed that there is "no health hazard" if the substance specific limit values of the lung damaging and toxic substances contained in the welding fume are not exceeded and the values of the carcinogenic substances are clearly below the relevant limit values. **Through the choice of effective ventilation measures the health hazard is reduced or even excluded**.

It is the basis for further evaluation at the workplace and for the choice of the relevant protective measures.

The Berufsgenossenschaft Holz und Metall (BGHM—German Social Accident Insurance Institution for the woodworking and metalworking industries) now offers on its website a software program for evaluating welding fume exposure by means of a hazard number.

The program is based on the model presented in the welding expert magazine "Welding and Cutting", edition 09/11 under the title Occupational health and safety regulations with regard to welding and assessment of the exposure to welding fumes and of their effect": The main emphasis is on the comparison between process, materials and workplace specific factors and not on hygienic-toxic evaluation. It becomes apparent why carrying respiratory equipment is a necessary supplementary protective measure for certain welding operations.

The software leads the user step by step into the description of the relevant workplace situation. The analysis is based on laboratory, practical measurements and other factors and provides a hazard number.

Its level provides the final decision determining what protective measure package needs to be assigned to the described workplace situation. The software also provides the user with tailor-made proposals for improvement. By repeating the calculations with the factors of the individual proposals, the user can find out from which of the proposals made he/she gets the most benefit.

References

1. Spiegel-Ciobanu, V. E. (2009). *Matrix zur Beurteilung der Schadstoffbelastung durch Schweißrauche, Band 3/2009: Aachener Berichte Fügetechnik* (Herausgeber Prof. Dr.-Ing. Reisgen). Shaker Verlag.
2. Spiegel-Ciobanu, V. E. *Welding activities with chromium and nickel alloyed filler and base metal, DGUV-Information 209-059 BGI 855E.*
3. Recommendations for hazard evaluation per the hazardous substances ordinance—tungsten intert gas welding (TIG welding). *DGUV Information* 213–712. https://publikationen.dguv.de/regelwerk/informationen/551/bg/bgia-empfehlungen-fuer-die-gefaehrdungsbeurteilung-nach-der-gefahrstoffverordnung-wolfram-inertgas.

International Standards

1. ISO 10882-1:2011: Health and safety in welding and allied processes—Sampling of airborne particles and gases in the operator's breathing zone—Part 1: Sampling of airborne particles.
2. ISO 10882-2:2000: Health and safety in welding and allied processes—Sampling of airborne particles and gases in the operator's breathing zone—Part 2: Sampling of gases.
3. ISO/DIS 15011-1:2015: Health and safety in welding and allied processes—Laboratory method for sampling fume and gases—Part 1: Determination of fume emission rate during arc welding and collection of fume for analysis.
4. ISO 15011-2:2009: Health and safety in welding and allied processes—Laboratory method for sampling fume and gases—Part 2: Determination of the emission rates of carbon monoxide (CO), carbon dioxide (CO_2), nitrogen monoxide (NO) and nitrogen dioxide (NO_2) during arc welding, cutting and gouging.
5. ISO 15011-3:2009: Health and safety in welding and allied processes—Laboratory method for sampling fume and gases—Part 3: Determination of ozone emission rate during arc welding.
6. ISO 15011-4:2006 + Amd 1:2008: Health and safety in welding and allied processes—Laboratory method for sampling fume and gases—Part 4: Fume data sheets.
7. ISO 15011-5:2011: Health and safety in welding and allied processes—Laboratory method for sampling fume and gases—Part 5: Identification of thermal-degradation products generated when welding or cutting through products composed wholly or partly of organic materials using pyrolysis-gas chromatography-mass spectrometry.

© International Institute of Welding 2020
V. E. Spiegel-Ciobanu et al., *Hazardous Substances in Welding and Allied Processes*, IIW Collection, https://doi.org/10.1007/978-3-030-36926-2

Printed in the United States
by Baker & Taylor Publisher Services